A Treatise on Good Robots

THE LEARNED SOCIETY OF PRAXIOLOGY

PRAXIOLOGY:
The International Annual of Practical Philosophy and Methodology Vol. 21

A Treatise on Good Robots

Praxiology, Volume 21

Krzysztof Tchoń
Wojciech W. Gasparski
editors

Routledge
Taylor & Francis Group
LONDON AND NEW YORK

First published 2014 by Transaction Publishers

2 Park Square, Milton Park, Abingdon, Oxfordshire OX14 4RN
711 Third Avenue, New York, NY 10017

Routledge is an imprint of the Taylor & Francis Group, an informa business

First issued in paperback 2017

Library of Congress Catalog Number: 2013021570

Library of Congress Cataloging-in-Publication Data

A treatise on good robots / Krzysztof Tchoń and Wojciech W. Gasparski, editors.
pages cm
Includes bibliographical references.
ISBN 978-1-4128-5285-2
1. Robotics—Human factors. 2. Robotics—Social aspects. 3. Robotics—Moral and ethical aspects. I. Tchoń, Krzysztof. II. Gasparski, Wojciech.
TJ211.49.T74 2014
629.8'92—dc23

2013021570

ISBN 13: 978-1-4128-5285-2 (hbk)
ISBN 13: 978-1-138-50731-9 (pbk)

The Learned Society of Praxiology expresses gratitude to the Ministry of Science and Higher Education for supporting the preparation of this volume.

On the cover: The social robot FLASH, built at the Wroclaw University of Technology, Poland. Photo courtesy of Jan Kedzierski.

In Memoriam

Prof. ZBIGNIEW J. GACKOWSKI, PhD
(1929–2013)

who supported and contributed to the series Praxiology:
The International Annual of Practical Philosophy and Methodology.

Contents

Part Three: Good Robots—Philosophical Perspective

Editorial

Wojciech W. Gasparski
Editor-in-chief

This volume is the latest one in a series proving the existence of a link between praxiology and technology. The previous volumes were *Design & Systems: General Applications of Methodology* (Collen and Gasparski 1995) and *Praxiology and the Philosophy of Technology* (Gasparski and Airaksinen 2008). One should also mention *The Roots of Praxiology: French Action Theory from Bourdeau and Espinas to Present Days* (Alexandre and Gasparski 2000) in which excerpts from a treatise by Alfred Victor Espinas, the discipline's originator, were published. Let us recall his idea:

> "Each of us, in fact, belongs to a social milieu, so to speak, to a world which takes it upon itself to interpret the rules and generally saves us the trouble of deciding 'what is done' and 'what is not done.' From that point of view, each social group is no less characterized by its arts than each species by its instincts. . . . Let us observe that what are in question here are not fine arts but useful arts. The Greeks called them 'technai' and we could give them the name of *techniques* to distinguish them from the arts which tend to produce aesthetic emotion. . . . It would . . . be most profitable if we could use that word, as the Greeks did, for conscious and well thought-out practices, to a certain extent opposed to simple practices, which develop spontaneously, prior to any analysis. For it is fully developed arts, not unconscious practices, which gave birth to the science we are dealing with, and generate Technology. Each of them implies a special technology, so that the sum of these partial studies naturally forms systematic general Technology. The word *practice* probably contains a wider meaning; it can easily be taken as a concrete substantive (practice, practices), it is suitable for all collective manifestations of the will, whether spontaneous or well thought-out. It provides an excellent term to refer to the science of that order of facts as a whole: Praxiology. . . . Both words Technology and Praxiology may be adopted, the latter (Praxiology) to

> refer to the most general part, the former (Technology) to refer to the slightly less general part in the same group of research." (Espinas 2000, 46–47)

Both Technology and Praxiology have made a gigantic leap since those words were published for the first time in the late nineteenth century. It is not necessary to present the history of technological change in full, for it is known and experienced in our everyday life. It suffices to mention that on December 17, 2003, we celebrated the hundredth anniversary of the very first flights of the brothers Orville and Wilbur Wright, which lasted: the first—twelve seconds over a distance of fifty-three meters, and another (the fourth in the same day)—fifty-nine seconds over a distance of just 260 meters, while now the world without aviation is unthinkable. Our practice is remarkably different than it was in the period called the *fin de siècle* of the times when Espinas wrote his book. Praxiology as a philosophical reflection on our practice, however, is less well known, so it is worth briefly outlining the history of inquiry in this field, particularly Polish praxiological studies, seeing as their outcomes have not been published in today's *lingua franca*, a fact noticed by Western philosophers:

> "Tadeusz Kotarbiński's ideas were related to our physicalism. He maintained conceptions which he called 'reism' and 'pansomatism', i.e., the conception that all names are names of things and that all objects are material things. Both Leśniewski and Kotarbiński had worked for many years on semantical problems. I expressed my regret that this comprehensive research work of Leśniewski and Kotarbiński was inaccessible to us and to most philosophers in the world, because it was published only in the Polish language, and I pointed out the need for an international language, especially for science." (Schilpp 1963, 30–31)

It was 1913, just one hundred years ago (yet another anniversary!) when the first praxiological collection, *Practical Essays: Issues of the Philosophy of Action*, composed of the papers written earlier by the young Polish philosopher Tadeusz Kotarbiński, was published in Polish under the title *Szkice praktyczne: Zagadnienia z filozofii czynu* (Kotarbiński 1913). Only one of the essays, "The Goal of an Act and the Task of the Agent," Kotarbiński's very first contribution to the discipline, presented in 1910, was translated into English and published thirty years ago in another collection, *Praxiological Studies: Polish Contributions to the Science of Efficient Action* (Gasparski and Pszczołowski 1983, 1–19). During the period of seventy years between the two dates, philosophy of action acquired its name of praxiology, after Espinas (1897), and achieved maturity expressed in Kotarbiński's opus magnum,[1] *Traktat o dobrej robocie* (A Treatise on Good Work, in Polish) (Kotarbiński 1955). The book was translated into English ten years later but published under a different title: *Praxiology: An Introduction to the Science of Efficient Action* (Kotarbiński 1965). Why was the title of the book changed? The publishers probably considered the original title to be insufficiently impressive from a marketing point of view or, quite the contrary, too strange for readers outside Poland. However, whatever the reason, it is important to refer to the original Polish title of the treatise, for it is crucial for understanding the title of the present volume.

Why? Because it is an obvious play on the relevant words in Polish: "robota" (work), "robotnik" (worker), and "robot" (artificial worker). The play on words and the reason is as follows: once we have a treatise on "dobra robota" (good work) written in the pre-cybernetic period, it is time now, because of Technology's development, when cybernetics and artificial intelligence are flourishing, to examine the similarities and differences between: (a) praxiology of human workers (including human knowledge workers (Drucker 1993)) and (b) praxiology of artificial workers, called "robots," that is, to sketch a tentative treatise on the possible praxiology of robots. The *Treatise on Good Robots* has been composed with such an intention.

There are quite a few different words in different languages for different kinds of human actions. The most popular is "work." Subjectively, "work" is perceived as something *difficult, hard,* or even *tough*. Because this negative connotation of work predominates, the Greek word "ponos" also means *suffering* and its Latin equivalent "labor" (earlier "labos") means *pain*. It is interesting to notice that "labor" traces its origin back to the Sanskrit word "rabhas," after which appeared the Gothic "arbeiths," Lithuanian "darbas," German "die Arbeit" accompanied by "Arebei" (pain), and finally the Slavic "rab" or "rob"—a slave, and in Polish "robotnik" (worker), that is, one whose activity is called "robota" (work) done mainly for somebody else, more often under pressure. Another etymological chain traces the origin back to the Sanskrit word "varg"—to set something in motion by force; from which comes the Greek word "ergon" as an outcome of one's work (i.e., work as a product), the German "werk," English "work," and Lithuanian "vargas"—a slave. The Romans translated "ergon" as "opus," the Italians created "operaio"—a worker, and many others—"operator" and "operation."

Work was painful and humiliating for centuries, and it was done by slaves.[2] The ancients valued relaxation, in Latin "schole," "otium," so they named activity or work through words opposed to those words, that is, "ascholia," "negotium." "Linguistic data do not offer material solid enough for one to answer in what conditions those words were formulated. We don't know if work was tough because of tiredness, or because of social inequalities and those who worked being held in contempt by those who did not need to work, benefitting from the efforts of others" (Pszczołowski 1966, 7–13). Tadeusz Kotarbiński wanted to liberate work ("robota") from this bad heritage by giving it, especially good work ("dobra robota"), the rank of general human activity *ergo* human action.[3] Action, according to Ludwig von Mises, an Austrian praxiologist, is the ultimate given of a human being (Mises 2007). This is why it is said "you will know them by their deeds."

The above offers an additional reason to compose a treatise on good robots under the praxiological umbrella. The reason is twofold: (1) Arnold Kaufmann, a French praxiologist, suggested *biopraxiology* as "the study, in living nature, of strategies of the species to defend, to maintain, or even to transform themselves" (Kaufmann 1968, 229) and (2) Carl Mitcham, an American technology philosopher, noticed that Espinas' and Kotarbiński's ideas "blend into what are now called systems theory, game theory, cybernetics" (Mitcham 1994, 33), so it is reasonable to investigate whether it would be possible to draft—paraphrasing Kaufmann's definition—a study, in the artificial nature of robots, of their strategies to defend, to maintain, or even to transform themselves while at the same time being friendly and cooperative to humans, that is, being not just "robots" but *good robots* which are able to do *good work*. It is, however, the human designers' responsibility to create such robots. "The perfect robot," writes Kotarbiński in his treatise, "would be an automaton which could handle different situations. An automaton with movements uniquely determined for every standard situation would not satisfy that condition, which is satisfied precisely by a robot which is an acting subject" (Kotarbiński 2000, 152), that is, a human being not reduced to a tool-like automaton.

> "The recognition that work is a basic human right and duty entails that labor-saving devices should only be welcome when they do not contribute to unemployment. In particular, automation should be adjusted to our basic needs and legitimate aspirations, not the other way round. Robots should only be employed to eliminate hazardous or extremely boring jobs, never to create unemployment." (Bunge 1989, 334)[4]

Anthropomorphization of robots would also be improper and risky.

> "The dilemma of technology is the dilemma of contemporary humanity. If we cannot come to grips with ourselves—not individually, but individually and socially, and politically—we *will* find ourselves in the Frankensteinian fate. Not because

> Frankenstein has come to life, but because we have failed to see that it is we who are still pushing the buttons which give him the semblance of life. The earth has been delivered to our care and we do not now dare avoid facing the critical ambiguity which belongs essentially to our situation." (Ihde 1983, 124–25)

Taking this into account, it seems that the praxiological values of good work, namely (in alphabetical order): accuracy, adroitness, correctness, courage, durability, efficacy, effectiveness, efficiency, handiness, maturity, prudence, purity, rationality, reliability, resourcefulness, serviceableness, simplicity, solidity, usefulness, (Kotarbiński 1966), are valid for both human actors and their proxies, that is, robots designed and made by competent humans for human needs. Praxiological values are the virtues of practical conduct. They, like Aristotelian virtues, should be relevant to particular practical situations and their contexts. Moreover, grammar of action, like any language grammar, fixes only formal norms of correctness. One may follow them for the good and the right or the bad and the wrong aims and means; therefore, the virtue of ethicality must not be ignored (Goujon and Dubreuil 2001) as a precondition of really effective, efficient, and ethical, therefore *good*—in all meanings of the term—robots and the *robota* (work) they perform.

* * *

It is with great sorrow that we report the passing of Professor Arne Collen, PhD (1942–2012) who supported and contributed to the series *Praxiology: The International Annual of Practical Philosophy and Methodology* and served on its International Advisory Board since 1998. He co-edited *Design & Systems: General Applications of Methodology*, 1995, Vol. 3 of the series, and was the author of the book *Systemic Change through Praxis and Inquiry*, 2003, Vol. 11 of the series. In his writings "he has examined Polish praxiology and shown its relevance and richness to developing a continental U.S. American pragmatic approach to research methodology for systemic change."[5]

Dr. Arne Collen was a long-standing member of the Executive Faculty at Saybrook University, San Francisco, and the Research and Teaching Faculty at the California School of Organizational Studies and the California School of Professional Psychology of Alliant International University, San Francisco Bay Campus. On his website, Arne Collen placed the following *Autobiographical Statement*:

> "Over the course of my life I have been drawn to multiple means of finding value in the arts, humanities, and sciences. Strong early influences of my mother, an artist, poet and educator, and my father, a physician devoted to computers and informatics

in medicine, have been of profound consequence in provoking a dynamic of tension, struggle, and eventual integration that has lasted a lifetime. I have been blessed with good fortune to have had a rural childhood bringing me into intimate contact with nature, a rich and diverse high school and university education, and many opportunities to pursue a career in education settings adjacent to mainstream universities, affording me study, travel, and professional development, thereby enabling all the contents found in my life-culminating web site. Although I have not devoted my life entirely to art, just as I have not done so in the humanities and sciences, I have found a rewarding and endlessly rich middle way in which my passions of life-long learning became manifest in my study of research methodologies for human inquiry across and among the arts, humanities, and sciences. If one had to put a label on it, the phrase would be research methodologist. I have found few companions taking this more challenging path, but it has not dissuaded me one iota from likely one of the most important decisions made early in my youth, now gratefully appreciated in my later years. The pursuit has enabled me to be artist, writer, and scientist, or to my way of thinking, an inquirer on a quest to know what one can about myself, others, the world, and all of life that makes one's existence meaningful and fulfilling. It is my sincerest desire that in sharing the contents of my web site in the form of an arts based autobiography, others may come to know me and find something of value too." (http://www.arnecollen.com/autobiographical-statement/)

Notes

1. See also (Gasparski 1983).
2. In many places of the contemporary world, despite technological changes, sweatshops still exist. The *Longman Dictionary of Business English* defines "sweated labour" as: "industrial workers who work very long hours in bad conditions for very low wages (in a place called a sweatshop)" (Adam 1989, 453).
3. ". . . action theorists, with the sole exception of the Polish school of praxiology, have systematically ignored work. Why may this be so: general lack of academic interest in important problems, or aristocratic contempt for praxis?" (Bunge 1989, 334).
4. See also (Forester and Morrison 1995).
5. From: About the Author, in; A. Collen, *Systemic Change through Praxis and Inquiry* (New Brunswick, NJ and London: Transaction Publishers, 2003), 304.

References

Adam, J. H. *Longman Dictionary of Business English.* New Edition. Burnt Mill, Harlow: Longman York Press, 1989.

Alexandre, V., and W. Gasparski. *The Roots of Praxiology: French Action Theory from Bourdeau and Espinas to Present Days.* New Brunswick, NJ and London: Transaction Publishers, 2000.

Bunge, M. *Treatise on Basic Philosophy, Vol. 8, Ethics: The Good and the Right.* Dordrecht: Reidel, 1989.

Collen, A. and W. Gasparski. *Design & Systems: General Applications of Methodology.* New Brunswick, NJ and London: Transaction Publishers, 1995.

Drucker, P. F. *Post-capitalist Society.* Oxford: Butterworth-Heinemann, 1993.

Espinas, A. V. *Les Origines de la technologie.* Paris: Alcan, 1897.

Forester, T., and P. Morrison. *Computer Ethics: Cautionary Tales and Ethical Dilemmas in Computing.* Cambridge, MA: The MIT Press, 1995.

Gasparski, W. "Tadeusz Kotarbiński's Methodology of the Practical Science and Its Influence." In *Research in Philosophy & Technology*, vol. 6, ed. P. T. Durbin, 93–106. Greenwich, Conn and London: Jai Press, 1983.
Gasparski, W., and T. Airaksinen. *Praxiology and the Philosophy of Technology.* New Brunswick, NJ and London: Transaction Publishers, 2008.
Gasparski, W., and T. Pszczołowski, eds. *Praxiological Studies: Polish Contributions to the Science of Efficient Action.* Boston, MA and Warszawa: Reidel-PWN, 1983.
Goujon, P., and B. H. Dubreuil, eds. *Technology and Ethics: A European Quest for Responsible Engineering.* Leuven: Peeters, 2001.
Ihde, D. "Technology, Utopia and Dystopia." In *Research in Philosophy & Technology*, vol. 6, ed. P. T. Durbin, 107–25. Greenwich, Conn and London: Jai Press, 1983.
Kaufmann, A. *The Science of Decision-Making: An Introduction to Praxeology.* London: Weidenfeld & Nicolson, 1968.
Kotarbiński, T. *Szkice praktyczne.* Warszawa, Kasa im: Mianowskiego Publ, 1913
Kotarbiński, T. *Traktat o dobrej robocie.* Wrocław: Ossolineum, 2000. 1st ed., Łódź, 1955.
Kotarbiński, T. *Praxiology: An Introduction to the Science of Efficient Action.* Oxford and Warszawa: Pergamon Press and PWN, 1965.
Mises, L. von. *Human Action: A Treatise on Economics.* Auburn, AL: Ludwig von Mises Institute, 2007.
Mitcham, C. *Thinking through Technology: The Path between Engineering and Philosophy.* Chicago, IL: The University of Chicago Press, 1994.
Pszczołowski, T. *Praca człowieka.* Warsaw: Wiedza Powszechna, 1966.
Schilpp, P., ed. *The Philosophy of Rudolf Carnap.* La Salle: Cambridge University Press, 1963.

Introduction

Krzysztof Tchoń
Wrocław University of Technology
Wrocław, Poland

> *"I can assemble anything that enters my head," said Microx. "But on the other hand not everything enters it . . . Whatever we create, we create from matter . . . and in it are contained all possibilities. . . . However, there are more possibilities in matter than in our heads, the thing to do, then, is to provide matter with a mouth, that it may tell us itself what else can be created from it, which would never cross our minds!"*
>
> *"A mouth is necessary," agreed Gigant. "But a mouth alone is not sufficient, for it expresses what the intellect within conceives. Therefore, we must not only give matter a mouth, but implant in it intelligence as well, and then it will surely tell us all its secrets!" (S. Lem 1977)*

This Treatise collects ten essays presenting a mosaic of perspectives on the Treatise leitmotif: the concept of a good robot. All of them have been rooted in robotics and artificial intelligence. For this reason, in order to introduce the reader into the universe of discourse of the Treatise, we shall make a bird's eye view of the developments and perspectives of these two scientific disciplines.

The year 1960 has been recognized as the birthday of *robotics* as a domain of science and engineering (Kumar 2010). In that year the first industrial robot was deployed in a General Motors plant in New Jersey. The father founders of industrial robotics were J. Engelberger and G. Devol (Menzel and D'Aluisio 2000). The industrial robots are programmable machines able to repeat indefinitely routine manufacturing operations. Tireless and precise hard workers, most often blind and deaf, industrial robots are totally dependent on their human programmers. For safety reasons the robots are separated from humans by security barriers. An older brother of robotics was *artificial intelligence* (AI) brought into being in 1956 at Dartmouth Conference, pioneered by McCarthy, Minsky, Newell, and Simon (Sabanovic, Milojevic, and Kaur 2012). The AI founders regarded robotics primarily as an attractive test bed of AI theories. For this objective, two AI laboratories were established, at MIT and at Stanford. It was Stanford where the well-known Stanford arm was developed as well as the first hand-eye system, and a mobile robot, called Stanford cart. A more sophisticated mobile robot, Shakey, was equipped with perception, navigation, and action planning skills (Shakey 2012). The first humanoid robot WABOT was built at Waseda University in Tokyo. A symbiosis between AI and robotics resulted in the now commonly accepted definition of robotics; *robotics is the intelligent connection of perception to action*, formulated by M. Brady in the MIT Artificial Intelligence Laboratory Memo entitled Artificial Intelligence and Robotics (Brady 1984). A robot endowed with perception (sensors) and action (effectors) capabilities, controlled by an on board computer (mind), could demonstrate autonomous, goal-oriented behavior, without direct human control. Moreover, the sensing, thinking, and expressive robots broke the robot-human barrier giving rise to service, assistive, and sociable robots, able to co-exist with humans and interact with them in a socially accepted way. By now the family of Waseda humanoids has grown substantially; Stanford University has become a renowned center of cooperative and human-centered robotics, while MIT has developed interactive humanoid and sociable robots (Menzel and D'Aluisio 2000; Breazeal 2002; Mertz 2010). Robots are becoming human assistants, caregivers, companions. To increase their believability, the robots appearance needs to be consistent with their cognitive capabilities; this design guideline prevents the occurrence of so-called *uncanny valley* (Mori 2012). A feature enhancing the relation of robot-human companionship is the ability of migration of the robot mind between various embodiments, for example, from the robot body to a computer or a mobile phone (Kheng et al. 2011). In a limit situation, the migration might involve the human mind uploaded into a robotic entity (Menzel and D'Aluisio 2000).

Inevitably, the creation of robots that co-exist and interact with humans has raised problems of responsibility of robotic designers for possible misbehavior of the robots. This kind of problems are addressed by *roboethics*, whose objective is making robotics to serve the advancement of humankind, and preventing its misuse (Veruggio and Operto 2008; see also Roboethics 2011). The roboethics is concerned with humans who build and use robots. However, the growing sophistication and autonomy of robots calls for establishing another kind of ethics, the ethics of conduct of robots as moral agents, called the *robot ethics*. In the face of increasing complexity of robots and possible unpredictability of their behavior even to the robot designers, it seems rational to incorporate some principles of moral behavior into the robots themselves. Asimov's Laws of Robotics are an example of such a set of moral commandments for robots.

A step further into the mutual permeance of the human and robot worlds or, more generally, biology and technology leads to the creation of *robotic humans* and *biological robots*. The former, called cyborgs are cybernetic organisms consisting of both organic and electronic parts. Most often this term refers to a human whose physical or cognitive capabilities have been enhanced by technology. The cyborgization of humankind, a sort of absorption of technology by biology, begins with various implants and prostheses, continues in our addiction to the Internet, benefits from the development of intimate brain-computer interfaces, etc. But, is it evolving toward the new species *robo sapiens*? (Menzel and D'Aluisio 2000). The opposite direction leads to the absorption of biology by technology resulting in the creation of biological robots. It has begun with bionics, the biologically inspired engineering which promoted the design of humanoid and android robots, and started to experiment with biohybrid robots whose control system incorporates whole living organisms or their parts like muscle or nervous tissue (Meyer and Guillot 2008).

In computer science, there is a well known *Moore's law* predicting the exponential growth of the number of transistors in an integrated circuit. A generalization of this law to the technology as a whole, and especially to artificial intelligence and robotics, will drive to emergence of the *technological singularity* (Vinge 1993), defined as an explosion of superhuman intelligence. Like the cosmological Big Bang, the technological singularity will give rise to the expansion of super-intelligent, transhuman robots, whose intelligence highly surpasses the intelligence of humans. The prophets of technological singularity have estimated its coming around the year 2050 (Kurzweil 2005; Moravec 2009). The process of self-improvement of the super-intelligent robots will produce generations of even more intelligent robots, rapidly accelerating the evolution of species, and concluding in replacing the homo sapiens by the robo sapiens. Can the transhumans be good?

This Treatise has been divided into three parts presenting the human, technological, and philosophical points of view. For the reader's convenience, its contents are concisely overviewed below.

The essay *How Good Robots Will Enhance Human Life*, opening Part One, consists of five sections written by K. Warwick, H. Shah, A. Vedder, P. Salvini, and E. Stradella, that take up the issue of good robots from diverse points of view. In the section entitled *From Science Fiction through Science Fact and Science Future* H. Shah argues that, despite gloomy visions of humanity exterminated by robots proliferated by science fiction literature and cinematography, the fact that the robots are going to surpass their human creators in many ways does not automatically imply that the robots become a threat to the humanity. Quite the contrary, the robotic technologies have already been and will enhance human life in all its aspects, beginning from advancing our knowledge due to the access to the Internet, through various robot assistive and companionship technologies up to the space robotics. The next section *Good Robot*, by A. Vedder, reflects on the meaning of the term *good* in the contexts of machines, humans, and animals. The author explains that in the machine world goodness is tantamount to functionality; goodness of a person is defined in ethical terms and means that the person's behavior is moral, utile or virtuous, whereas a good animal behaves well. The section is concluded with a suggestion that a good robot should: function well and reliably, be morally good, and not harm or offend the humans. The section *Good Humans Make Good Robots*, by P. Salvini, approaches the question of what a good robot is from the opposite side and defines a bad robot. A robot can be called as bad when does not have good intentions toward humans, does not perform well, is dangerous, or can be used in military operations. Bad robots exist either because they have been badly designed or badly used by humans. This yields a praxiological conclusion that in order to have good robots it is necessary to have good robot designers and good robot users. Making good robots is then to be preceded by deep interdisciplinary research of the possible side effects of emerging technologies, both contemporary as well as prospective. The last two sections of this chapter refer to two products of the advanced technology that can be called robotic humans and biological robots. The text by E. Stradella, entitled *Technology "Inside" Humans: Good Robots in the Perspective of Fundamental Rights*, deals with the problem of *goodness* of cyborgs or robotic humans that are humans whose physiological processes are to a certain extent controlled by electronic devices. The problem is studied via examples, such as neuro-prostheses, technological grafts on human body, cyborg technologies for military intelligence collection, and the Deep Brain Simulation. A characterization of a good cyborg states that a good cyborg should enhance the human and improve his/her physical and cognitive abilities to understand the world and other people, while good robotics is a personalization of technology, that should be dedicated to the blossoming of the mankind. The issue of biological robots is addressed in a study of K. Warwick on *Morality in Biological Robots*. A biological robot is a robot whose control system (the brain), instead of silicon chips, uses cultures of biological neurons. Such a culture can be grown *in vitro* and interfaced with the robot body by an array of electrodes, such that the culture of neurons works as a robot controller that is able to learn. Experiments are reported involving a mobile robot controlled by this type of biological brain. The questions stated in conclusion refer to the legal, cognitive, and ethical status of a biological robotic brain whose number of neurons approaches or even exceeds the number of neurons in the human brain.

The next essay in this Part, by C. Urdiales, U. Cortés, and R. Annicchiarico, entitled *The Good Assistive Robot for Elder Care* concentrates on socially assistive robots and specifically on robotic caregivers. The text has been based on a ten-year authors' research work on European Projects, devoted to the development of assistive technology. The objective of assistive technology is to provide *technical devices that could eliminate, improve, or compensate for functional limitations* of the human. The social significance of assistive technology in the face of aging human population and simultaneous decreasing availability of human caregivers is inappreciable. A potential applicability of robotic caregivers includes care of the elderly as well as of people with physical or cognitive disabilities in their home environments. Three groups of such robots have been distinguished: assistant robots, companion robots, and assistive robots. The assistant robots are autonomous robots, capable of accomplishing some general physical tasks to help the human, and able to learn and adapt to the human during the cohabitation. Some of them may serve merely as a communication means with the external world. Robotic companions engage in multi-modal social interactions. Specifically, they can recognize human emotions and respond by displaying robotic emotions. In this way the companion robots develop and maintain a companion relationship with humans. The assistive robots deliver a physical support to humans that need rehabilitation or just assistance in their daily life activities, like transportation and mobility. A peculiarity of assistive robots is the request of somewhat auto-restricted assistance:

the cared-for person should be assisted as little as possible in order not to lose his/her own skills. This request determines the functioning of the robot control system. The autonomy of the assistive robot should be restricted and in a sense be shared with the human. The shared autonomy results in the shared robot control that means that the control is handed over from the human to the robot and vice versa in such a way that at any time instant the control rests on either of these parts, the other being excluded. Alternatively, the shared control may take the form of the collaborative control combining simultaneously the human and the robot control actions, and not excluding the human completely, even in situations, where the robot control needs to dominate, for example, for safety reasons. A collaborative control system has been implemented on the robotic wheelchair CARMEN developed and designed by the authors, and subject to experimental tests whose organization, progress, and results have been described. It is concluded that a good assistive robot should be included into an ecological framework embracing humans and their environment. The good robot should be multi-tasked, interactive, adaptive, evolvable, reliable, attractive, affordable, and not stigmatizing. The robot's behavior is subject to ethical assessment. The good robot ought to assist, never to rule.

Part One is concluded with the contribution of Z. Gackowski entitled *Unorthodox Thoughts about Good Robots Viewed Pragmatically and Praxiologically* that represents a praxiological standpoint. The author begins by arguing that instead of good we should rather speak of well-designed or high quality robots. These terms apply to robot's actions, so the author passes to an analysis and explanation of three fundamental criteria of assessment of human actions, referred to as the praxiological triad: ethicality, effectiveness, and efficiency. These criteria can be applied to all human controlled operations, including the robotic operations since robots are controlled by humans either directly (as in the case of remote control) or indirectly (autonomous robots are after all human designs). In his pursuit of clarifying what constitutes the high quality robots, the author performs a crushing critique of the Asimov's Laws of Robotics. The quality of robots is put in the context of the quality of operations, again defined in praxiological terms of ethicality, effectiveness, and efficiency, and of four categories of users differing in their ways of the quality assessment: organizational users, individual users, individual consumers, and environmental stakeholders. As a result of assuming the praxiological way of thinking, it is concluded that a good, that is, a high quality robot should be: recognizable, relevant, pragmatically meaningful, significantly material, on-site and on-time available, reliable, effectively usable, and effectively useful when engaged in operations. It is a challenge for humanity to prevent the appearance of evil robots, the robots that are not good.

Part Two begins with an essay, *Towards Application-Oriented Navigation*, written by A. Borkowski, that sketches a developmental perspective of mobile robotics. The main message of this text is that mobile robotics has reached enough maturity to enable mobile robots to drive off robotics research laboratories, and to navigate toward practical applications. In this context the author addresses the most fundamental problem of mobile robotics that of building maps of the environment, and argues that this problem can actually be replaced by the maintenance of maps. It is proposed that the mobile robot navigation inside buildings relies on the building documentation represented in the form of the Building Information Model. Outside buildings the robot can use the data provided by Google maps, and satellite or aerial photographs, in accordance with the standard of the Geographic Information System. Given the map, the robot task reduces to its maintenance and updating. In order to move efficiently, the mobile robot needs to localize itself, that is, to determine its position in the environment. To this objective the author is recommending the topological or mixed topological/metric localization. For localization inside buildings the robot may use either vision systems or visible labels, like UPC = Universal Product Code and QR Code = Quick Response Code, or radio detectors and RFID (Radio-Frequency IDentification) tags, deployed appropriately. The localization outside buildings can be based on GPS (Global Positioning System) and local landmarks. A motion of the mobile robot is subject to advance planning. The path-planning problem consists in defining an admissible path for the robot, leading the robot in the obstacle collision-free way from the starting to the terminal point. It is suggested that inside buildings the path planning can rely on the topological or hybrid map and involve a number of motion primitives. The path planning for terrain robots moving on roads can be based on the path planners applied in automotive navigation systems. Off road path planning may use the diffusion method. Generally, the incorporation of ready-to-use, pre-planned routes is deemed to increase the efficiency of the mobile robot planning system. In conclusion, a good mobile robot can be thought of as extra-laboratory, efficient, robust, and affordable.

The following essay, *Locomotion Mode Transition Control of Multi-Locomotion Robot*, authored by T. Fukuda, T. Aoyama, Y. Hasegawa, K. Sekiyama, and Q. Huang, resumes the problem of robots assistive to humans in more diversified environments than just the home conditions. The authors argue that such robots, referred to as life-supporting, are distinguished by their mobility, multitasking, and intelligence. In order to provide high mobility, their locomotion should be biologically inspired, that is, employ various locomotion modes known from the animal world. Special attention has been paid to three of them: biped walking, quadruped walking, and brachiation. The first is recommended when the ground is rigid, freeing hands to perform manipulation tasks. The quadruped walk, being more stable, allows moving in rough terrain or in narrow passages, while the brachiation abilities are desirable when moving on the ground is hardly possible whatsoever. The idea of incorporating many locomotion modes in a robot has been implemented in a prototype Multi-Locomotion Robot Gorilla, designed by the authors. The robot control system whose objective is to guarantee the robot locomotion stability is capable of switching between biped and quadruped walking, depending on external or internal data. The external data come from exteroceptors, and allow the robot to estimate whether it is moving on a flat ground or approaching a steep slope. In the former case, the robot accomplishes the biped walk, in the latter the quadruped walk is chosen, switching when the estimated terrain slope puts at risk the locomotion stability. A similar transition between these two locomotion modes can be triggered by internal data combined into an estimate of probability of falling down. Experiments using external and internal data for the locomotion stabilization have been reported. The locomotion control system is open for inclusion of additional locomotion modes, such as brachiation and ladder climbing. It may be concluded that a good biologically inspired assistive robot is to be a multi-locomotive robot.

A subsequent essay, *Maintenance Robot: A History and Reflections of a Japanese Engineering Scientist on Fukushima*, by H. Yoshikawa presents reflections on the necessity of establishing a general theory of maintenance, and developing a maintenance technology, from the perspective of the Fukushima nuclear power plant accident in 2011. The idea of creating a science of maintenance was proclaimed by this author in a visionary paper published in 1984. A couple of years earlier a prototype of the maintenance robot MOOTY was built at the Tokyo University under this author's leadership. In the text, the state of the art in the maintenance theory and technology is assessed, and measures proposed that should be taken in order to make a significant progress in the maintenance. A strong motivation for undertaking this effort flows from the sense of responsibility of scientists and engineers in the face of disasters like that at Fukushima.

In turn, the maintenance theory should stimulate an advancement of the maintenance technology. A fundamental task in this context consists of the introduction of automatization and robotization into the maintenance technology. The lessons learned from the Fukushima disaster indicate a need of undertaking basic research on maintenance, both on the physical as well as systemic level, and a demand of developing the maintenance technology, including the robotic technology. Also, a change of attitude of scientists and engineers is mandatory, toward removing barriers between compartmentalized science and industry, improving communication of scientists and engineers with society and politicians, and shifting the interests of researchers from individual, egoistic subjects to the problems of primary significance for the society. A good robot appears then as a maintenance robot serving the society.

The closing Part Two essay, *Technological Singularity: A Path towards Artificial Intellects and Unpredictable Future*, by J. Brzezinski, is considering potential benefits and threats of the rapid growth of emerging technologies, such as artificial intelligence, neuroscience, nanotechnologies, and robotics, in the face of impending technological singularity. The author concentrates on a number of research trends in artificial intelligence heralding the appearance of artificial intellects, such as the brain research, the perspective of quantum, and DNA-based computers as well as the progress in real-time image recognition and the natural language processing. All these achievements may be regarded as signs of the time when intelligence will emerge from the activities of artificial brains. In contrast, advances of intelligent technologies will inevitably accelerate the process of cyborgization of the mankind. On the one hand, the progress in brain research and neuroscience fosters advances in biologically inspired artificial intelligence; however on the other hand, it is inherently connected with formidable threats, if we think, for example, of the mind-reading and brain-scanning technologies. The social, political, ethical, and even religious consequences of the technological singularity are given special attention. The following human attitudes toward artificial intellects have been distinguished: strong opposition and ban on intelligent technologies, similar to that of the mass destruction weapons, embracement as a chance of overcoming human biological limitations, and approval as a possibility of creating God-like entities and immortal minds.

In the essay initializing Part Three of the Treatise, entitled *Robot Desires*, T. Airaksinen makes an excursion into a philosophy of technology. He begins with somewhat provocative mentioning of the Shannon's Ultimate Machine whose behavior could be interpreted as driven by a desire of committing suicide. The author distinguishes basic needs as extensional entities, essential to the survival in the external world. These basic needs are common to both humans as well as robots. Contrary to the basic needs, the desires are understood as intentional states of mind that involve imagination, thought, wish, hope, want, intention, choice, experience, welfare or happiness, and value. Human desires feature all these ingredients, while in the robot we can find merely the thought; therefore, desires can be attributed exclusively to humans. The next problem addressed in this text refers to the transhumanist doctrine and the relationship between humans and cyborgs. The author's view can be summarized as the ascertainment of invariance of the human and the robot condition with respect to technology. Specifically, this means that neither designing more and more sophisticated robots will transform them into humans nor the cyborgization of humans by equipping them with advanced technological aids will deprive them of the status of human beings. It is argued that the human psyche is a fixed point of both these lines of development. The robots and the cyborgs will remain separate, similar to the undead and the alive on the opposite sides of the uncanny valley. Things look different, if a wide concept of cyborgs is assumed that claims that the human's transformation into cyborgs results from our embedding into a cybernetic reality. In the social world the humans behave humanly; however, within the cybernetic world our behavior resembles programmed entities with very basic needs. In pursuing the thought of the role of contemporary technology, the Heidegger's categories of resource and frame have been employed. The technology has a dual face, on one hand it transforms the world into the resource, and on the other it frames us by defining our desires, hopes, and plans. The final conclusion is rather pessimistic: the robots do not have and will never have desires, but our desires, the desires of technologically framed humans become increasingly to resemble the robotic ones.

The philosophical thread is continued in the essay, *From Good Robots to Useful Intelligent Agents*, written by G. Nalepa. The author is exploring the relationship between artificial intelligence and robotics. He begins with a vision of a universal, intelligent humanoid robot, then states that the real existing robots essentially depart from the vision, and concentrates on the robot's intellect regarded as an intelligent agent, extracted from its robotic embodiment. In this way, a good robot is replaced by a good agent defined as an agent that is useful to its user. The goodness of an agent can be achieved by means of the tools and techniques of artificial intelligence. Significant attention has been paid to symbolic knowledge representation and processing, rule-based reasoning, machine learning, user interfaces, distributed architectures, and context awareness. Among possible lines of development of the intelligent agents the author points to machine emotions and computational creativity.

The Treatise is concluded by the essay, *On the Significance of "Responsibility" in Technological Thinking*, written by L. Tondl. This text belongs to the *philosophy of technology*, a philosophical current pioneered half a century ago by T. Kotarbiński, that in our times has been known under the name of technology assessment, meaning an integration of knowledge and values. Tondl defines responsibility as an ethically stimulated duty of rational actors whose purpose is to select humanly acceptable objectives and means of action, paying attention to its results, benefits as well as potential risks. This understanding of responsibility applies in particular to such human actions like designing technical artifacts. A specific area of application of the technology assessment is the rapidly expanding field of information/communication technology and the artificial intelligence. The author focuses on the Janus-faced character of responsibility, its significance in any rational, that is, goal-oriented action, as well as on its major attributes. He stresses that in all phases of the technical work, it is the human actor who bears responsibility, not only for its creation and use, but also for its potential abuse. Beyond professional competencies, a responsible actor is identified by personal qualifications that include ethical standards, ability of considering potential risks, openness to elicit other opinions and options, and readiness for conducting interdisciplinary consultations and negotiations. The recognition of the problem of responsibility in the area of robotics and artificial intelligence has been summarized in the old Latin recommendation *Quidquid agis, prudenter agas et respice finem*.

References

Brady, M. *Artificial Intelligence and Robotics*. MIT Artificial Intelligence Laboratory. Memo 756, 1984.

Breazeal, C. L. *Designing Sociable Robots*. Cambridge, MA: The MIT Press, 2002.

Kheng, K. L., D. S. Syrdal, K. Dautenhahn, K. Arent, Ł. Małek, and B. Kreczmer. "Companion Migration—Initial Participants' Feedback from a Video-Based Prototyping Study." In *Mixed Reality and Human-Robot Interaction*, ed. Xiangyu Wang, 135–51. Dordrecht: Springer, 2011.

Kurzweil, R. *The Singularity is Near.* New York: The Viking Press, 2005.

Lem, S. *How Microx and Gigant Made the Universe Expand, Mortal Engines*. New York: The Seabury Press, 1977.

Menzel, P., and F. D'Aluisio. *Robo Sapiens: Evolution of a New Species*. Cambridge, MA: The MIT Press, 2000.

Mertz. http://www.designboom.com/technology/mertz-robot/, 2010.
Meyer, J. A., and A. Guillot. "Biologically Inspired Robots." In *Springer Handbook of Robotics,* ed. B. Siciliano and O. Khatib, 1395–422. Berlin: Springer, 2008.
Moravec, H. "Rise of the Robots—The Future of Artificial Intelligence." 2009. http://www.scientificamerican.com/article.cfm?id=rise-of-the-robots.
Mori, M. "The Uncanny Valley." *IEEE Robotics & Automation Magazine* 19, no. 2 (2012): 98–100.
Roboethics. *IEEE Robotics & Automation Magazine*, special issue 18, no. 1 (2011).
Sabanovic, S., S. Milojevic, and J. Kaur. "John McCarthy." *IEEE Robotics & Automation Magazine* 19, no. 4 (2012): 99–106.
Shakey. http://www.sri.com/works/timeline/shakey-robot, 2012.
Veruggio, G., and F. Operto. "Roboethics: Social and Ethical Implications of Robotics." In *Springer Handbook of Robotics*, ed. B. Siciliano and O. Khatib, 1501–24. Berlin: Springer, 2008.
Vinge, V. "The Coming Technological Singularity: How to survive in the Post-Human Era." VISION-21 Symposium, NASA Lewis Research Center and Ohio Aerospace Institute, March 30–31, 1993. http://mindstalk.net/vinge/vinge-sing.html.
Kumar, V., "50 Years of Robotics." *IEEE Robotics & Automation Magazine*, 17, no. 3 (2010): 8.

Part One

Good Robots—Human Perspective

How Good Robots Will Enhance Human Life

Kevin Warwick and Huma Shah
School of Systems Engineering
University of Reading, UK

Anton Vedder
Tilburg Institute for Law, Technology, and Society
Tilburg University, Netherlands

Elettra Stradella, University of Pisa, Italy

Pericle Salvini, Scuola Superior Sant'Anna, Pisa, Italy

1. Introduction

The authors of this chapter, Kevin Warwick (KW), Huma Shah (HS), Anton Vedder (AV), Pericle Salvini (PS), and Elettra Stradella (ES) are research partners in a European Consortium entitled RoboLaw. The aim of their research is to investigate the ways in which emerging technologies in the field of (bio-) robotics (e.g., bionics, neural interfaces, and nanotechnologies) have a bearing on the content, meaning, and setting of the law. However, this necessarily involves both the philosophical and technological study of robots and what it means for the robots to exist as good and moral entities, and how they will enhance their human compatriots.

This chapter is wide ranging with each author's section a standalone point considering robots in science fiction, including in the groundbreaking cinematic *Metropolis*; the notion of *good* robots; the need for *good designers* to create good robots; the technology *inside* good robots; and *morality* in biological robots. Robots and robotic technologies are already enhancing human life, as in the case of cyborg artist Neil Harbisson with an implant and eyeborg device (see section by ES). Among the many realistic aims, future robots will be involved from the domestic mundane task of washing our dishes, to the healthcare and companionship of the elderly, to exoskeletons to enable the wheelchair bound to stand and walk, to assisting astronauts in space exploration for the good of humanity (CompanionAble 2012).

2. From Science Fiction through Science Fact and Science Future [HS]

The notion of an "artificial human" as slave, companion, and then overlord is centuries old. In the industrial era of the early twentieth century the Czech Capek brothers, Josef (painter and writer) and Karel (playwright) bequeathed the terms *automat* and *robota* to the world. But it is from the mind of Thea von Harbou, in the screenplay of her husband Fritz Lang's 1927 silent era black and white epic *Metropolis*, described as "one of the most iconic and influential films of all time" and the first film "to be included on the UNESCO Memory of the World register of essential cultural artifacts" (Metropolis1927.com), that a stark image of cyborgs and machines emerged (see Figure 1, below).

Figure 1
From Left: Robot Creation; Robot Close-Up; Robot with Rotwang the Inventor; Robot Centre

In *Metropolis*, a technicized "mother of all cities," the future is portrayed with workers, not walking but shuffling along their cramped haunt, exhausted from their shift in the machine halls deep underground while the liberated elite of the city enjoy a decadent lifestyle, as high above the lethargic workers as the latter's subterranean living quarters are below (see Figure 2). The mechanical female, who cannot be distinguished from the mortal Maria, the heroine who gives the underground workers hope by prophesying a better future to come, is given Maria's physical attributes and sexuality by the mad inventor Rotwang. The cyborg is programmed to wreak havoc and is charged with deceiving the workers to stir up trouble among them (YouTube 2012).

Figure 2
Metropolis: Elite City; Underground Workers

From Lang's 1927 masterpiece many movies since have embraced this ill-wind portrayal of machines and robots, from *2001: A Space Odyssey* with its murderous HAL computer, to *Blade Runner*'s Nexus 6 Replicants seeking extended existence, to the killer cyborgs in the *Terminator* movies. With the advent of artificial intelligence the myth of the Frankenstein monster turning on its creator has now "entered the realm of the possible" (Lehman-Wilzig 1981: p. 442). There is a recurring theme in such films: robots are essentially untrustworthy, inherently malfeasant, and will become killers, so humanity is doomed. People are taking this seriously. Echoing Eliezer Yudkowsky et al.'s assessment of catastrophic risk from intelligences created by humans (2010), in April 2012, Cambridge University, UK set up a centre to study "existential risk" (CSER 2012). The trio behind the Cambridge quest (philosopher Huw Price, ex-Royal Society President Martin Rees, and co-Skype founder Jaan Tallinn) view developments in human technology as possible "extinction-level risk" to the human species (ibid). Price and Tallinn (2012) believe that the "caravan of narrow AI" driving specialist task-driven technology "moves steadily forward"; the philosopher and the technologist claim that machines have surpassed human performance in "chess, trivia games, flying, driving, financial trading, face, speech and handwriting recognition." But why should this advancement be seen in a negative light? Could it be because we accept, open a broadsheet newspaper on any day to find news of, humans blighting other humans' existence around the world as so common that a corollary follows that what we humans create will necessarily finish us, the unethical, the illogical, and the immoral species off?

Taking a step back we notice that human offspring often surpass their parents in many ways, yet how many kids bump their mother and father off? Human Rights Watch warns of killer robots (Human Rights Watch 2012), but unless they are programmed to, why should robots kill off humanity? Taking a step forward, could it not be the case that intelligent robots will show us humans the way to be humane; or to be more humane more often? Taking a bird's eye view of the *current step* we find technology and robots advancing human life, bringing in the good and extracting the difficulties faced by some in their day-to-day living. Computers and the Internet connection allow us to have the "mother of all libraries" at our fingertips any time of the day or night, all the time enhancing our knowledge, increasing our intelligence. Smart mobile phones now incorporate high specification cameras, video capability, as well as many other functions including listening to our favorite tunes, making this product the first thing many of us see and use in the morning and last thing at night. Parking-assist technology in road vehicles has made it easier and easier to park cars in tight and difficult spots while reducing damage to its body. Advanced driver-assist technology is in the pipeline and self-driving cars "which conceive everything around them and make decisions about every aspect of driving" (Thrun 2011) will be seen on our roads in the near future helping to prevent traffic accidents, reducing injury and casualty. Wearable technologies, *exoskeletons*, assist the wheelchair bound to stand and walk; chip implants lift the depression from patients like Brandy Ellis, to help them feel human and be a part of the society (BBC 2012). Domestic robot technologies mow our lawns and hoover around the home. Robots are being developed to help astronauts work in space (NASA's 2012, *Robonaut*), to care for and companionate the elderly (CompanionAble's *Hector*), while some humanoids are destined to fulfill desires as sexbots for the paying sex market (Levy 2012). Robot prostitutes could reduce the violence and inhumanity suffered by many humans forced into this profession; whether it would be ethical to create "child robots" to assuage the desires of pedophiles, thus mitigating the threat to vulnerable children, is a debate for another day.

These are but a few examples of robots and robotic technologies that already enhance, and will enhance human life. The robots of the future could be tiny and in our head allowing us a cinema or concert experience anytime we should choose; they could be personalized and wearable technologies that transport us safely along the roads, or fly us to Cancun or Hawaii on a cold winter's day in Europe. If we can imagine "good" then there is no reason for future robots not to be designed to "be good" and further improve the quality of human life.

3. Good Robot [AV]

What can the notion of "good" mean in the expression "good robot"? One might think that this question arises merely because currently a robot is still a phenomenon seldom seen. Since robots are still rare, people will find it difficult to explain what a good robot looks like. However, even in the presence of an abundance of robots it might still be difficult to decide what a good robot would amount to. "Good" is a primarily a formal notion referring to several other properties and dimensions of objects and even to norms and classifications in the minds of the users of the vocabulary. Notions such as these have been called "essentially contested" (Gallie 1956), meaning that their definitions will go on to provoke controversies, since the norms and classifications at issue will understandably be the subject of ongoing debates.

Although any hope of final and conclusive answers to our initial question is therefore probably in vain, we might nonetheless kindle reflection on the issue by looking at the use of the notion of good in combination with a machine, a human being, and an animal respectively.

What is a good vacuum cleaner? A good vacuum cleaner is undoubtedly a machine that cleans well, helps you to get rid of dirt and dust in the first place. The primary criterion for goodness of a machine seems to build on its functionality: the degree to which it does what it was made for, in this case: cleaning. Additional—but not unimportant—reasons for calling it good may be found in side-aspects of the good functioning of the machine: that it cleans well without using too much energy, neither in terms of power—electricity—nor in terms of the physical involvement of the person who operates the machine, that it does not cause any accidental damage, and that it manifests some constancy with regard to its performance and operating life. Both the primary and the additional criteria for its goodness have to do with *expectations* of people with regard to this type of machine. Of course, these expectations are somehow informed by the current state of art in the relevant techniques and technologies, or even by de facto and formal technical standards, but we can safely assume that in day-to-day conversations they are based on implicit or explicit expectations, which a specific machine is or is not living up to.

What is a good man or a good woman? It is difficult to think of other definitions of good men and women than definitions in moral terms. Again, the exact definition might depend on the moral stand one takes: A Kantian may say that a good man or woman is a person generally acting in accordance with moral principles. A Utilitarian might argue that a good man or woman is a person whose behavior and life style brings about the greatest utility for most people affected. An advocate of virtue ethics again may contend that a good man or woman is a person who shows excellence of character, attitudes, and actions. This latter view of the good person may come closest to the way in which we define good when referring to machines: often advocates of virtue theories refer to a unique characteristic or purpose of human beings, for example, being rational or being a social being, when stipulating the specific excellences of character, attitudes, and action.

What is a good dog? It may seem a bit odd to refer to a dog as "a good dog." It reminds us of the good machine, on the one hand, and of the good person, on the other. But it is difficult for us to think of animals in general as having a purpose or having a function. Of course, animals kept with a special purpose, such as giving milk, towing, or guiding blind people, might be referred to as a good milk cow, a good workhorse, or a good seeing-eye dog, respectively. They will nonetheless only somewhat awkwardly be referred to as good animals or as a good cow, good horse, or good dog. The association with the goodness of persons is also awkward. With regard to people, "good" connotes the morality of an actor, and it is certainly odd to think of animals as moral actors. Perhaps the closest we could get to meaningful use would be by defining "good dog" in terms of behaving well or of minimal decency (not barking at us or biting us) and, again, living up to our expectation.

Having seen this: What would a good robot be? Perhaps we might distinguish the following possibilities:

- A robot that is functioning well, doing what it is expected to do, with a reasonable life span, and not requiring too much attention or maintenance.
- A robot that is morally good (as far as its behavior and its consequences are concerned, and, depending on its degree of autonomy, its "attitudes," "dispositions," or "intentions").
- A robot that does not harm or offend us or scare us.

4. Good Humans Make Good Robots [PS]

Another way to understand what a good robot is would be to look at its opposite, namely: bad robots. What is a bad robot? We propose four simple ways in which a robot can be considered bad:

1. a robot whose intentions are not good toward human beings, for example, like Terminator,
2. a robot that does not perform well, for example, a cleaning robot that does not clean efficiently,
3. a robot dangerous for human beings, animals, the environment, and other objects, for example, a lawn mower robot not designed according to safety standards,
4. a robot used in military operations, for example, a drone.

With the exception of the world of science fiction, bad robots as described in point no. 1 do not (currently?) exist. On the contrary, there exist robots which are considered bad according to point nos. 2 to 4. This is basically the case of robots *badly designed*, which also includes robots that are badly programmed, manufactured, assembled, maintained, etc. or *badly used*, that is, robots that are used with bad intentions by human beings. A *badly designed* robot can cause several problems. It can be a robot that does not perform well the task that it is supposed to do (e.g., the robot does not clean well), or that it is not sufficiently robust (e.g., the robot performs well, but it breaks down often), or that it has not been designed according to safety standards (e.g., the robot is dangerous for human beings, animals, or other things), or that it is not usable, that is, it is not designed according to ergonomics or human factor criteria (e.g., a robot toy too high to be used by children). A *badly used* robot can mean that the robot has been used improperly. Damages resulting from improper use of technology can be divided into malicious (e.g., an exoskeleton which, instead of being used for enhancing a person's physical capabilities, is used for teleoperating human beings), and not voluntary (e.g., an industrial robot arm used as a kitchen appliance). Furthermore, a *badly used* robot may be the result of a bad design, as in the case of robots that are designed to carry out a task which is considered ethically bad, illegal, or socially unacceptable (e.g., military robots or robots replacing human beings in "sensitive" tasks, such as elderly or children care, education, social relation).

In order to have good robots we need to have good designers and good users first. However, even the best intentions, the correct use of machines, as well as a flawless design, are not enough. As a matter of fact, technology is not neutral: "we shape our tools and thereafter they shape us" (Culkin in (Stearn 1968)). From the invention of the wheel to the Internet, technology has been modifying the ways in which we relate to each other and the environment we inhabit. So to make good robots it is always necessary to have good interdisciplinary research into the potential side effects of new technologies. As Isaac Asimov once pointed out: "It is change, continuing change, inevitable change, that is the dominant factor in society today. No sensible decision can be made any longer without taking into account not only the world as it is, but the world as it will be . . . This, in turn, means that our statesmen, our businessmen, our everyman must take on a science fictional way of thinking" (Asimov 1978).

5. Technology "Inside" Humans: Good Robots in the Perspective of Fundamental Rights [ES]

Could the goodness of robots have a juridical significance? A reply to this question shall be given considering a specific application of robotic technologies such as cyborg technology. Indeed, cyborgs are the "human robots," that is, humans characterized by certain physiological processes that are at least partially controlled by mechanical or electronic devices becoming actually integrated in the human body. This means individuating them as legal subjects for all intents and purposes. As a matter of fact, whilst robots' legal subjectivity is quite far to be reasonably taken into consideration, because of their still limited autonomy, cyborgs are a concrete phenomenon in which what is good and what is bad can clearly emerge. Actually, cyborgs are the specular expression of what will be analyzed in the following paragraph: "robotic humans" versus "biological robots."

The goodness (or not) of robotics in human being can be explored via four main examples. The first example regards neuro-prostheses that raise the issue of the definition of responsibilities (for damages to others and to one's own body), in the occurrence of an interaction between humans and technology, where the latter is integrated in the human being. The issue mainly deals with the damages caused by robots, in this case the "human robots" or cyborgs. This issue is being deeply investigated by legal scholars (Santosuosso 2012), but has not been solved yet. The case of Christian Kandlbauer, an Austrian man who was the first in Europe to wear an innovative high-tech artificial arm and who died in a car accident, illustrates the controversial definition of what is good and what is bad in cyberlimbs. Although in the accident it was impossible to tell whether the event was caused by problems with Kandlbauer's prosthetic arms (and in particular the bionic arm), anyway the causal nexus between a malfunctioning (not merely mechanic) of the bionic arm and the damage requires a rethinking of the traditional categories of civil liability because of the necessary evaluation of the degree of autonomy in the interaction between technology and human behavior.

Another challenging field is the technological "graft" on human body realized for philosophical or artistic purposes, to demonstrate the intrinsic boundaries of the human body and the potentialities of technology in development that surpass and definitively replace it. In this field, the goodness of the cyborg shall be evaluated through the capacity of technology to effectively become an expressive instrument, respecting the limits to the freedom of having the use of one's own body, that are fixed in a critical (and necessarily changeable) concept of human dignity.

The example of Stelarc (see Figure 3), the well-known technological performer artist, gives us an idea of the controversial relationship between the freedom of bodily expression and the goodness of the intervention by technology on the human body. As Stelarc observes, his research follows the assumption that "To be human is to be augmented, extended and enhanced by technology" (Goodall 2005). In this approach, technology "constructs our human nature," and technological "grafts" do not aim at making visible the originary technicity of the human (Galvan and

Figure 3
Stelarc with An Extra Ear

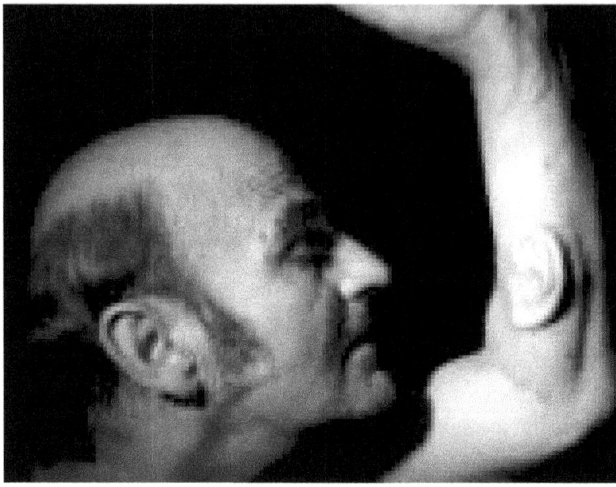

Luppicini 2012). but at demonstrating "how different developments in technology (robotics, the Internet, virtual reality systems, prosthetics, medical instruments and procedures) alter our conception of the human and of the human body" (Massumi 2002).

The case of Neil Harbisson makes all these suggestions concrete and legally relevant. In 2004 Harbisson, who asked for renewing his passport, did not succeed because the British passport office would not allow him to appear in the photo with electronic equipment on his head: Harbisson was born with achromatopsia and only a cyber-implant allowed him to perceive colors by memorizing the frequencies related to each color (see Figure 4).

Harbisson wrote back to the office asserting that the eyeborg should be considered part of his body as he had become a cyborg (see Figure 4). In the end, Harbisson's device was included in the photo for his passport, someway confirming his statement according to which the union between the software of the device and his brain converted him into a cyborg (Miah and Rich 2008). This was the legal recognition of the cyborg essence, and its legitimacy (or goodness, if ethically considered).

Third example is that of the military use of cyborg technologies in the field of intelligence collection (it is the case of the escargot-spy, published in the Journal of American Science in 2012, or the well-known scarab-spy whose patent has been published in 2011 in the American Journal of

Image 4
Neil Harbisson with An Eyeborg

Micromechanics and Microengineering) that poses the question about the international legitimacy of these instruments. The dual-use and the problematic application of Robotics to the military field have already been discussed, but some different features emerge with regards to the intelligence collection activities, since it could be particularly hard to determine the legal consequences derived from the conducts undertaken by these "strange" subjects.

Finally, we consider Deep Brain Stimulation (DBS) and the interference of technology on human free will and desire. This case strictly connects the goodness of cyborg with the right to an informed consent that maybe should differ with respect to existing procedures, in order to protect the fundamental right to health and self-determination of people (patients) that are able to condition a multiple series of future behaviors and choices with a single action. Today, the main application of DBS regards patients with Parkinson's disease (Perlmutter and Mink 2006), but more sensitive situations concern people affected by psychiatric disorders and compulsive eating. Technology is good depending on the degree of compliance to the effective needs of patients and to the guarantee of some fundamental rights such as health and autonomy. An alteration of the self-determination processes in the individual indeed is able to affect these rights.

A few remarks to conclude this section:

- the good cyborg enhances the human being in their own nature and characters, giving an answer to human intrinsic aspiration to improving their capacity to know and understand the world and to be connected with other individuals;
- the tiresome debate between transhumanists/posthumanists and neoluddists is nullified if we focus on the human being with her fundamental human rights and on the good robots intended as what does not contrast with rights and liberties protected in the constitutional heritage of the national and international charters;
- traditionally, robots are perceived as more and more acceptable when they tend to plausibly resemble humans or do not look like them at all (it is the well-known Uncanny Valley phenomenon), but good robotics is the "personalization" of technology in the sense of its vocation to the fulfillment and flourishing of humans.

6. Morality in Biological Robots [KW]

When one first thinks about robots, what commonly springs to mind are autonomous robots whose main controlling elements/brains are computer based in the sense of silicon microprocessors (Bekey 2005). Issues then arise with regard to the identity of the robot and whether or not it is a conscious being. It is now possible however for robots to have their own biological brain, made up of thirty million highly connected biological neurons, either originating from rats or humans. In this section, we look at the method of origination of such a robot and consider some of the implications in terms of morality and ethics.

Neurons can be grown under laboratory conditions on an array of non-invasive electrodes. An experimental robot body can then move around in a defined area purely under the control of such a network/brain and the effects of the brain, controlling the body, can be witnessed.

Typically, culturing networks of brain cells in vitro commences by separating neurons obtained from rodent cortical tissue. They are then grown in a chamber, in which they can be provided with suitable environmental conditions (e.g., temperature) and nutrients. An array of electrodes embedded in the base of the chamber (a Multi Electrode Array) acts as an electrical interface to/from the culture. The neurons in such cultures spontaneously connect, communicate, and develop.

The culture is grown in a glass specimen chamber lined with a planar 8x8 Multi Electrode Array (MEA) which can be used for real-time recordings (see Figure 5). It is possible to separate the firings of small groups of neurons by monitoring the output signal on the electrodes. In this way a picture of the global activity of the network can be formed. It is also possible to electrically stimulate the culture via any of the electrodes to induce neural activity. The MEA therefore forms a bi-directional interface to the cultured neurons (Chiappalone et al. 2007; Warwick et al. 2010).

When it is sufficiently developed, the culture can be coupled to its physical robot body. Sensory data fed back from the robot is subsequently delivered to the culture, thereby closing the robot-culture loop. The electrochemical activity of the culture is sampled and this is used as an input to the robot's wheels. Meanwhile the robot's (ultrasonic) sensor readings are converted into stimulation signals received by the culture.

Figure 5
(a) An MEA Showing the Electrodes, (b) Electrodes in the Centre of the MEA Seen Under An Optical Microscope, (c) An MEA at x40 Magnification, Showing Neuronal Cells in Close Proximity to An Electrode

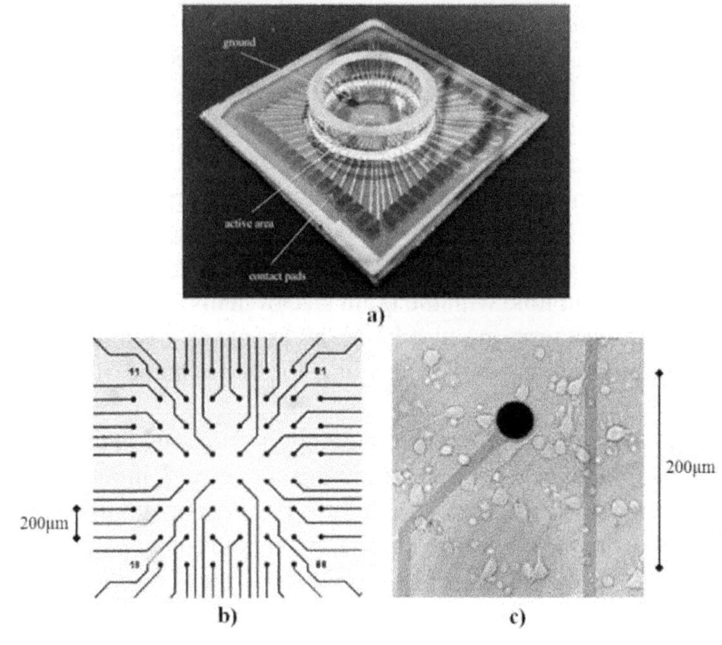

An existing neuronal pathway is identified by searching for strong relationships between pairs of electrodes. This is initially employed to control the robot body—for example, if the ultrasonic sensor is active and we wish the response to cause the robot to turn away from the object being located ultrasonically in order to keep moving.

For experimentation purposes, the robot moves forward until it reaches a wall, at which point the front sonar value decreases below a threshold, triggering a stimulating pulse. If the responding/output electrode registers activity, the robot turns to avoid the wall. In experiments, the robot turns spontaneously whenever activity is registered on the response electrode.

As an overall control element for direction and wall avoidance, the cultured network acts as the sole decision making entity within the overall feedback loop. Clearly one important aspect involves neuronal pathway changes, with respect to time, in the culture.

Learning and memory investigations are generally at an early stage. However, the robot improves its performance over time in terms of its wall-avoidance ability in the sense that neuronal pathways that bring about a satisfactory action strengthen purely through the process of being habitually performed—learning due to habit. Learning by reinforcement—rewarding good actions and punishing the bad ones is an investigative research effort.

On many occasions the culture responds as expected, on other occasions it does not, and in some cases it provides a motor signal when it is not expected to do so. But does it "intentionally" make a different decision to the one we would have expected? We cannot tell.

In terms of robotics, it has been shown by this research that a robot can successfully have a biological brain to make all its "decisions." The 100,000-neuron size is due to the limitations of the experimentation described. Well in tune with the brain size of many other animals, three-dimensional structures are also being investigated, realizing a figure of thirty million neurons—not yet reaching the one hundred billion neurons of a human brain.

This area of research is expanding rapidly. Not only is the number of cultured neurons increasing, but the range of sensory input is being expanded to include audio and visual. Such richness of stimulation will no doubt have a dramatic effect on culture development. The potential of such systems, including the range of tasks they can deal with, also means that its physical body can take on different forms. There is no reason that the body could not be a two-legged walking robot, with a rotating head.

Understanding neuronal activity becomes more difficult as the culture size increases. In fact, the present 100,000-neuron cultures are already far too complex for us to gain an overall insight of the neuronal activity. When they are grown to sizes of thirty million neurons and beyond, clearly the problem is significantly magnified. It seems realistic to assume that such cultures will become larger, potentially growing into sizes of billions of neurons. On top of this, the nature of the neurons is being diversified. Human neurons are also being cultured realizing a robot with a human neuron brain. If this brain then consists of billions of neurons, many social and ethical questions will need to be asked (Warwick 2010).

For example—if the robot brain has roughly the same number of human neurons as a typical human brain then could/should it have similar rights as humans? Also—what if such creatures had far more human neurons than in a typical human brain—for example, a million times more—would they make all future decisions rather than regular humans? But will it choose to be a good robot; indeed, will "good" have meaning to the robot?

7. Conclusions

In this chapter we raised some issues that concern how good robots will enhance human life: meaning of good, the need for good designers for good robots, the technologies inside good robots, and whether biological robots can be moral. Robotic technologies already assist humans in their life: Neil Harbisson's eyeborg device helps the color-blind artist to "see" color; deep brain implants are used to treat Parkinson's disease and are also considered in psychological treatments. Domestic robots help keep our homes clean. In the near future, robots will care for and companionate the increasing elderly population here on earth, while in space they will assist astronauts on the international space station, and will go onto explore deep space and send back information helping humans to make the big decisions to enhance and prolong humanity.

8. Acknowledgment

The authors wish to acknowledge, with gratitude, the financial support of the European Commission for the RoboLaw project (ID: 289092) under the call FP7 – SCIENCE – IN – SOCIETY – 2011 – 1.

Figure Credits

1. and 2. From *Metropolis* movie website: http://metropolis1927.com/#gallery.
3. From Stelarc image gallery on Stelarc website: http://stelarc.org/?catID=20290.
4. Of Neil Harbisson: Dan Wilton / Red Bulletin—http://harbisson.com/.

References

Asimov, I. "My own view." In *The Encyclopedia of Science Fiction*, ed. Robert Holdstock. 5, New York: St. Martins' Press, 1978.

BBC, "Brandy Ellis: Being a Cyborg Restored my Humanity." Radio 4 Digital Human, 2012. http://www.bbc.co.uk/programmes/p010p0ll.

Bekey, G. *Autonomous Robots.* Cambridge, MA: MIT Press, 2005

Chiappalone, M., A. Vato, L. Berdondini, M. Koudelka-Hep, and S. Martinoia. "Network Dynamics and Synchronous Activity in Cultured Cortical Neurons." *International Journal of Neural Systems* 17 (2007): 87–103.

CompanionAble, "2012, Integrated Cognitive Assistive & Domotic Companion: Robotic Systems for Ability and Security." *EU FP 7 Funded Collaborative Project*, 2012. http://www.companionable.net/.

CSER. *The Cambridge Project for Existential Risk.* Cambridge University, UK. 2012. http://cser.org/.

Gallie, W. B. "Essentially Contested Concepts." *Proceedings of the Aristotelian Society* 56 (1956): 167–98.

Galvan, J. M., and R. Luppicini. "The Humanity of the Human Body: Is Homo Cybersapiens a New Species?" *International Journal on Technoethics,* 3, 2 (2012): 1-8.

Goodall, J. "'The Will To Evolve' Stelarc: The Monograph." Ed. Marquard Smith. MIT Press, 2005.

Human Rights Watch. "Losing Humanity: The Case against Killer Robots." HRW & IHRC Report, 2012. http://www.hrw.org/reports/2012/11/19/losing-humanity-0.

Lehman-Wilzig, S. N. "Frankenstein Unbound: Towards a Legal Definition of Artificial Intelligence." *Futures—IPC Business Press* 30 (1981): 442–57.

Levy, D. "The Ethics of Robot Prostitutes." In *Robot Ethics: The Ethical and Social Implications of Robotics*, ed. P. Lin, K. Abney, and G. A. Bekey, 223–31. Cambridge, MA: MIT Press, 2012.

Massumi, B. *Parables for the Virtual: Movement, Affect, Sensation.* Durham, NC: Duke University Press, 2002.

Metropolis1927.com. "*Metropolis.*" 2010. http://metropolis1927.com/#home.

Miah, A., and E. Rich. *The Medicalization of Cyberspace.* New York: Routledge, 2008.

NASA. "Robonaut RS2." 2012. http://robonaut.jsc.nasa.gov/default.asp.

Perlmutter, J. S., and J. W. Mink. "Deep Brain Stimulation." *Annual Review of Neuroscience* 29 (2006): 229–57.

Price, H., and J. Tallinn. "Artificial Intelligence—Can We Keep It in the Box?—The Conversation: Latest Ideas and Research." 2012. http://theconversation.edu.au/artificial-intelligence-can-we-keep-it-in-the-box-8541.

Santosuosso, A., C. Boscarato, and F. Caroleo. "Robot e Diritto: Una Prima Ricognizione." *La Nuova Giurisprudenza Civile Commentata* II, no. 7/8 (2012): 494.

Stearn, G.E., ed. *McLuhan Hot & Cool*. Harmondsworth: Penguin, 1968.

Thrun, S. "Google's Driverless Car." TED 2011, 2011. Retrieved from: http://www.ted.com/talks/sebastian_thrun_google_s_driverless_car.html.

Warwick, K. "Implications and Consequences of Robots with Biological Brains." *Ethics and Information Technology* 12, no. 3 (2010): 223–34.

Warwick, K., S. Nasuto, V. Becerra, and B. Whalley. "Experiments with an In-Vitro Robot Brain." Chapter in *Instinctive Computing*, ed. Y. Cai, Lecture Notes in Artificial Intelligence 5987 (2010): 1–15.

YouTube. "Metropolis." 2012. http://www.youtube.com/watch?v=ekL3VBD00W0.

Yudkowsky, E., C. Shulman, A. Salamon, R. Nelson, S. Kaas, S. Rayhawk, Z. Davis, and T. McCabe. *Reducing Long-Term Catastrophic Risks from Artificial Intelligence*. San Francisco, CA: The Singularity Institute, 2010.

The Good Assistive Robot for Elder Care

Cristina Urdiales
ISIS Group, Universidad de Málaga, Spain

Roberta Annicchiarico
Fondazione Santa Lucia, Rome, Italy

Ulises Cortés
KEMLg Group
Universitat Politècnica de Catalunya, Barcelona, Spain

1. Abstract

Taking the appropriate care of older adults is among the major ethical, technological, social, and economic challenges currently faced by aging populations. The main aim of most proposals made until now is to give the *right* amount of support, thus allowing elders to safely live in a semi-autonomous way at their preferred environments as long as possible. Many of these solutions are based on the assumption of the existence of a *good* companion (sociable/assistive) robot.

Even if those robots do exist, there are a lot of factors like age, gender, education, experience with technology, and so forth that make this technology difficult to be widely accepted by needy elders. From our point of view, *adaptability* is a key factor for acceptance as well as the users' experience with technology.

We present our ideas showcasing a good assistive robot for elder's care, focusing on enhancing technology acceptance by means of adaptation to the user's specifics.

2. Introduction

Last century demographic changes have accelerated population aging, which, in turn, has had an impact on elders' care as many more individuals need, at different levels, (almost) continuous caregiving. There are currently approximately forty-five million people in Europe who report a long-standing health problem or disability (Applica CESEP Alphametrics 2007). Further, the aging population means that more people will have to live with some sort of disability. Recent figures from the 2008-based national population projections EUROPOP20082 show that the share of people aged sixty-five years or over in the total population is projected to increase from 17.1 percent to 30.0 percent and the number is projected to rise from 84.6 million in 2008 to 151.5 million in 2060. This means there will be more than 80 million people with several needs for care by 2050.

It is estimated that the cost of this care will range from 5 percent to 20 percent of GDP by 2025. This part of the population is growing rapidly world-wide; intelligent robotics has the potential to: (i) provide unique assistive technologies to improve Quality of Life (QoL) and (ii) serve as a tool for principled, analytical study of human behavior.

To enable robots to perform socially assistive roles, the first step is to allow them not only to physically interact but also to communicate with their *users*. Users may be a larger group of individuals, including the elderly themselves, as clinicians, social services, and/or caregivers that are empowered to have access to the information about the elderly's performance provided by the robot. This information may be used to characterize the kinds of interaction between elders and robots in a given environment and the types of support those robots ought to provide.

The aim of this essay is to establish the conditions for *good* robots to assist people and improve human performance in daily and task-related activities, focusing in particular on populations with special needs, including those convalescing from trauma, rehabilitating from cognitive and/or physical injury, aging in place or in managed care, and suffering from developmental or other social, cognitive, or physical disabilities. In particular, we are interested in support mobility in this target population.

The basic assumption in our research effort is that the target population by using these supporting devices could be self-dependent enough as to autonomously live in their community, staying at home as long as possible with maximum safety and comfort; this possibility would increase their quality of life and, at the same time, delay their institutionalization.

At least in part, how well these elders live and their abilities to maintain independent life styles will depend on their health and the degree to which they have remained totally able or frail or disabled. This in turn, will depend, to at least some extent, on how well the artificial and built environments, in which they live, conform to their needs and their age-related losses in abilities and somatic integrity (Cortés et al. 2010).

Our point of view is based on the work performed on developing intelligent assistive tools along the last ten years in creating various assistive robots as *CARMEN, SPHERIK*, and *i-Walker* in EU-funded project *SHARE-it* (Supported Human Autonomy 2007), and the rehabilitative version of *i-Walker* in the EU-funded projects *IDONTFALL* (Integrated Prevention and Detection Solutions 2012) and *FATE* (Fall Detector for the Elder 2012).

3. Assistive Technologies and Disability

Advanced information and communications technologies—artificial intelligence, image processing, natural language processing, and robotics—can be used to help elders and their caregivers cope with their daily living. To date (2013), many of the breakthroughs in advanced and autonomous technology, such assistive robotics, have only begun to offer real—but limited—social and economic benefits.

The great interest in the treatment of disability has been reflected in the use of technological tools more or less sophisticated, the so-called Assistive Technologies (AT) (Kamel Boulos et al. 2009; Cortés et al. 2010). The AT can be defined as *technical devices that can eliminate or improve or compensate for functional limitations*. These tools are able to support people with disabilities, while making more effective and efficient interaction of the subjects themselves with their surroundings, physical and social.

Currently, disability is defined as difficulty or dependency in carrying out activities necessary for independent living, including roles, tasks needed for self-care and household chores, and other activities important for a person's quality of life (Fried et al. 2004). The natural course of frailty is progressive, increasing the risk of comorbidity and disability over time. In fact, findings support the hypothesis that frailty causes disability, independent of clinical and subclinical diseases. The syndrome of frailty may be a physiologic precursor and etiologic factor in disability, due to its central features of weakness, decreased endurance, and slowed performance (Fried et al. 2001).

Different groups of elders are characterized by large variations in health, well-being, disability, and health care needs. Assistive Technologies' (AT)' major purpose is represented by the possibility of assisting people—elderly and/or disabled—to stay at or to get back to their home. Thus a target population—in terms of functional, social, and clinical features—has to be individuated so that a number of individuals as large as possible can achieve benefits as effectively as possible.

Robotics play a key role in providing smart assistance so that devices adapt to the user's needs and environment. Next section presents a brief review of robotics in elder care.

4. A Brief Taxonomy of Robots in Elder Care

Related investigations of ageing people and their real needs are not new, but they have not obtained the expected impact. Some reasons for this failure may be: study fragmentation, lack of critical mass, lack of vision of the relevant industry, and a missing political signal of favorable conditions for implementation and coordination on a large scale (Branch and Jette 1992; Broekens et al. 2009). In addition, the situation is not characterized by a lack of technologies, but by a lack of *complete* solutions, which should be easy to handle, easy to install, reliable, and affordable for the targeted people as well as the social safety systems (Fall detector for The Elder 2012).

Tapus et al. identified (Tapus et al. 2007) the following application domains of socially assistive robotics: (a) care of the elderly, (b) care of individuals with physical recovery/rehabilitation and training needs, and (c) care of individuals with cognitive and social disabilities. There have been many proposals to use robots for assistance at home during the last decade. Although all of them fit the Ambient Assisted Living (AAL) paradigm, they may belong to three different groups depending on their main purpose.

The so-called *assistant robots* are typically autonomous agents that help the user with minor tasks, like carrying food, drinks, and medication (Hagras et al. 2004; Cesta et al. 2005; Saffotti and Broxvall 2005). These robots are always task-oriented, and typically connected to the environment so they make decisions depending not only on their onboard sensors but also on the input instance of all ambient sensors, as well as any known user's agenda and habits that may be provided either a priori or learnt-through use. Since it is usually necessary to conciliate different behaviors, targets, and constraints, agent architectures are often used to deal with the system complexity. It is interesting to note that these robots usually adapt to living habits, but are usually not specifically designed for a given disability profile, nor treatment.

A special category in assistant robots is *socialization robots* (Cauchi et al. 2012; Granata and Bidaud 2012; Tiberio et al. 2012). Socialization robots are mobile agents, usually without manipulation capabilities. Their goal is to promote social inclusion by linking users to people and events outside their homes. Consequently, they typically hold a monitor to provide easy access to online services, ranging from videoconference or participation in social networks to cognitive reminders or activity monitoring. It is important to note that these robots are not expected to engage the user in any activity by themselves: they simply act as a mobile gateway to connect to other person(s) at the user's preferred environment. Nevertheless, as most assistant robots, they support voice-based interaction for simplicity.

Companion robots conform to the second group of these agents. Robot companions are expected to respond emotionally to situations experienced and to interactions with humans (Artificial Emotion). Robo-pets like Aibo ERS series or Ugobe Pleo have already been commercialized with limited success, and there is also a commercial therapeutic robot named Paro that has been used in clinical experiments. Paro, as similar robots like iCub, is based on learnt emotive expression in the behavior of the robot for human-machine interactions. Reported effects on persons affected by dementia include stress reduction, improved interaction, and socialization. It needs to be noted that, in most cases, it is not necessary for the robot to understand emotions at all, that is, build an artificial personality, but just to learn the adequate response to the user's instance. This process usually consists of categorizing facial expressions (e.g., anger, fear, distress/sadness, enjoyment/happiness, disgust, surprise, and so forth . . .) and additional factors like voice pitch and tone, proximity, body motion, etc. Then, the robot may trigger tentative behaviors and evaluate the response of the person engaged in interaction (prompted learning) or try to mimic what a person does in response to one of these situations (on-the-fly learning). More complex systems even try to create an artificial personality to drive social interaction between the user and the robot (Breazeal and Scassellati 1998).

Finally, *assistive robots* are typically used for physical support by people with special needs, both during rehabilitation processes or simply to carry their Activities of Daily Living (ADL). Rehabilitation robots are task-oriented and not necessarily mobile. We focus on mobile assistive robots, because they are the most critical: they must be used by people with special needs without supervision in dynamic environments for unspecified tasks.

4.1. Assistive Robots and Mobility

Transportation and mobility are closely linked to independence, well being, and quality of life. A critical factor in an older person's ability to function independently is mobility, the ability to move without assistance (Applica CESEP Alphametrics 2007; Espeland et al. 2007). Older people who lose mobility are less likely to remain in the community, have higher rates of morbidity and mortality, have more hospitalizations, and experience a poorer quality of life (Branch and Jette 1992).

In fact, mobility in addition to representing one of the most important basic ADL as defined in the Barthel Index (Mahoney and Barthel 1965) is also directly involved in many basic and instrumental ADL activities like transferring, grooming, toileting, housekeeping, etc. Everyday assistive robots typically support mobility and are shaped like canes, walkers, or power wheelchairs, depending on the needs of the target population. It is important to point out that, unlike other systems, these robots are usually expected to assist the user as little as possible, so people have to make some effort to prevent loss of residual skills. Hence, an important part of these systems includes interfaces and estimation of how much help is needed.

Robotic devices are similar to autonomous robots, regarding input instance (e.g., controller, encoders, range sensor) and navigation algorithms. However, physicians advise that the user's Intervention Level needs to be kept as high as possible, so these systems usually rely on shared control architectures. Shared control may range from safeguarded navigation (Miller 1998)—the wheelchair stops or takes over in case of danger to—in extreme—full autonomous navigation. Most commercial systems, though, rely on a basic set of frequently used primitives like AvoidObstacle, FollowWall, and PassDoorway to assist in difficult maneuvers (Connell and Viola 1990; Simpson et al. 1998). These behaviors can be either manually or automatically selected. Alternatively, other systems rely on combining human control and robot output in a continuous way. Hence, users are forced to contribute to control even in situations they are not comfortable with. The *SHARE-it* approach relies on giving the users just the amount of assistance to avoid loss of residual capabilities and at the same time enhance the feeling of autonomy of assisted people (Urdiales et al. 2009).

This work focuses on assistive robots, specifically on robot wheelchairs. Our goal is to evaluate the response of all agents involved in the process of using such a robot such as ADL-users, clinicians, engineers, and the robot itself—via an appropriate set of metrics and questionnaires. Ultimately, we aim at checking if shared control can efficiently cope with faced challenges, improve motivation, and achieve acceptance in our target population, consisting of persons with mild to severe physical and/or cognitive disabilities.

Agent autonomy can be defined as the ability to make independent choices. Specifically, an autonomous robot should be capable of performing a task in an unstructured environment without continuous human guidance. However, assistive robots specifically require intervention from a person for task achieving. Situations where machines and persons cooperate to achieve a common goal fall within the field of shared autonomy (Aigner and McCarragher 2000; Bruemmer et al. 2005). Shared autonomy algorithms can be categorized in terms of how much the robot does into i) safeguarded operation and ii) shared control. In the first case, mobile robots can be totally controlled by humans, but in some cases the robot may override their commands in case of imminent danger (Morris et al. 2003) or when human control is not adequate (Parikh et al. 2004; 2005). Shared control implies a more active robot: given a situation that the user has trouble to cope with, like door crossing, U-turn, etc., (Bruemmer et al. 2005) the user may choose to yield control on the robot. Alternatively, the robot might be able to detect certain situations and take control itself. It can be noted that, in general, in most shared autonomy systems control is either in the hands of the user or the robot at any given instance. However, there are algorithms that allow both the human and the robot to contribute simultaneously to wheelchair control rather than passing it over. Collaborative control may be considered a special type of shared control where it is necessary to have both robot and human providing commands at the same time. It is up to each specific algorithm to determine how those commands are combined into an emergent one. The main advantage of collaborative control from the system point of view is that discontinuities in the control function are drastically reduced. However, it is even more interesting to note that, from a medical point of view, the user needs to contribute to control all the time, even in situations he/she may find difficult. The *SHARE-it* approach achieves this goal by weighting the robot and human standalone commands at each situation by their respective local efficiency: the better the user performs, the less help he gets (Urdiales et al. 2009). Specifically, we measure local efficiency for any given command as the average between smoothness (η_{sm}), directness (η_{tl}), and safety (η_{sf}) (Figure 1). Smoothness depends on the angle between the current heading and the command direction (α_{dif}). Directness depends on the angle between the direction to the target and the command direction (α_{dest}). Safety depends on the angle between the repulsion vector to nearby obstacles and the command direction (α_{min}).

Figure 2 shows how collaborative control works. Robot commands in our work are provided by a purely reactive Potential Fields Approach (PFA). The user has been simulated by tying the joystick so that it's always pushed ahead. Thus, we achieve a *dummy* user capable of producing a single command: move ahead. The bars below the wheelchair picture show smoothness, directness, safety, and global efficiency both for the dummy user and collaborative control (top and bottom bar for each efficiency factor, respectively). It can be observed that in this point of the trajectory our dummy user has a high smoothness and directness—as expected, since it is not changing directions anytime, but a low safety, since it is moving toward an obstacle. All in all, the user's efficiency is higher than collaborative efficiency, but prevents the wheelchair from crashing into furniture. On the right of these bars, we can observe both PFA and dummy user's commands and the collaborative control emergent command, which is an average of both, since η_{sm} and η_{tl} almost compensate η_{sf} at this point.

Figure 1
Forces Involved in Collaborative Control Calculation for a Command Vector V

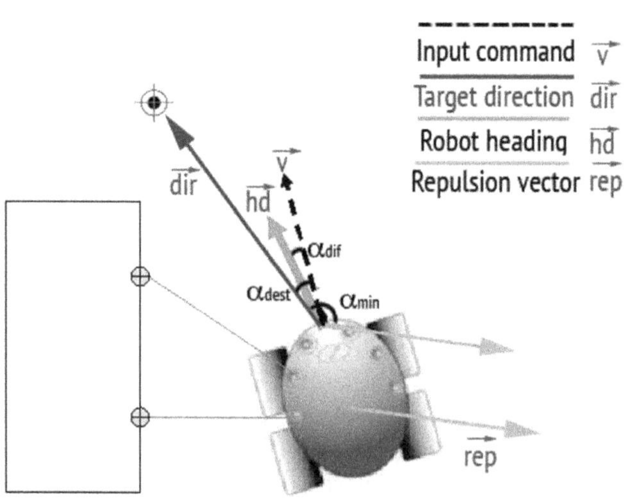

Figure 2
Collaborative Control Based on PFA for a Dummy User

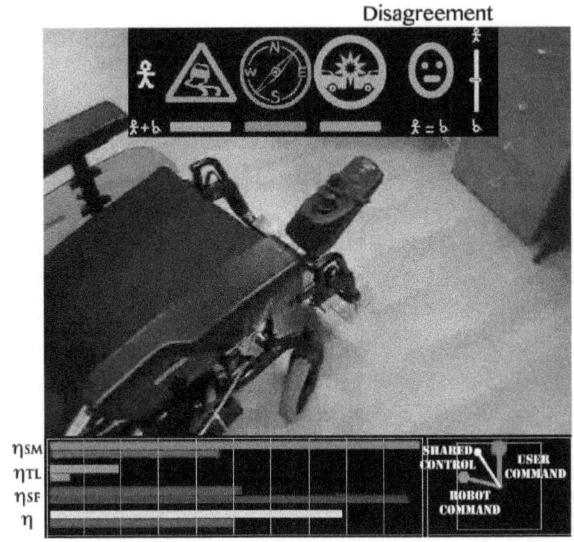

Closer to the obstacle, η_{sf} would be much lower and the vehicle would steer faster. On top of the picture, there is an iconic representation of all three efficiency factors for user and collaborative control, as well as a measure of the Disagreement, which represents how similar user and emergent commands are. The slider on the right of the Disagreement loosely represents how much control the user retains at each time instant. In this case, as commented, it is approximately a 50 percent.

It is necessary to note that it is not always possible, nor necessarily advisable to achieve a local efficiency equal to one. Some efficiency factors are opposite in presence of obstacles, like keeping far from obstacles and trying to turn as little as possible. Also, the layout of the environment may make it necessary not to head to the goal at all times. Similarly, in order to move through narrow places like doors, safety efficiency might be low sometimes. Emergent commands will be closer to humans whenever the person is driving better than the robot, and also that he/she always contributes to emergent motion anyway.

5. Scenario and Test Procedure

Our tests with collaborative control were conducted over a period of three years at Fondazione Santa Lucia (FSL), a rehabilitation hospital in Rome. More than seventy in-patients volunteered to drive a robotic wheelchair named *CARMEN* (Collaborative Assistive Robot for Mobility Enhancement) in a home-like environment.

This environment, *Casa* Agevole[1] is a *60 m*^{2} fully furnished, standard-compliant test house built within the FSL complex.

Tests were supervised by a medical team that provided in-patient medical data and took care of all medical protocols. All tests were approved by FSL Ethical Committee and performed with the volunteers' informed consent.

Regarding the eligibility criteria for the volunteers, we used a battery of neuropsychological tests, including Mini Mental State Examination (MMSE), fifteen-words (verbal episodic memory)—prose recall (verbal episodic memory), Rey's figure, Psychosocial Impact of Assistive Devices Scale (PIADS), among others, to identify the volunteers' cognitive levels. We were interested in measuring the following items:

- Memory functions
- Perceptual functions: visuo-spatial abilities
- Attention functions

Figure 3
CARMEN wheelchair

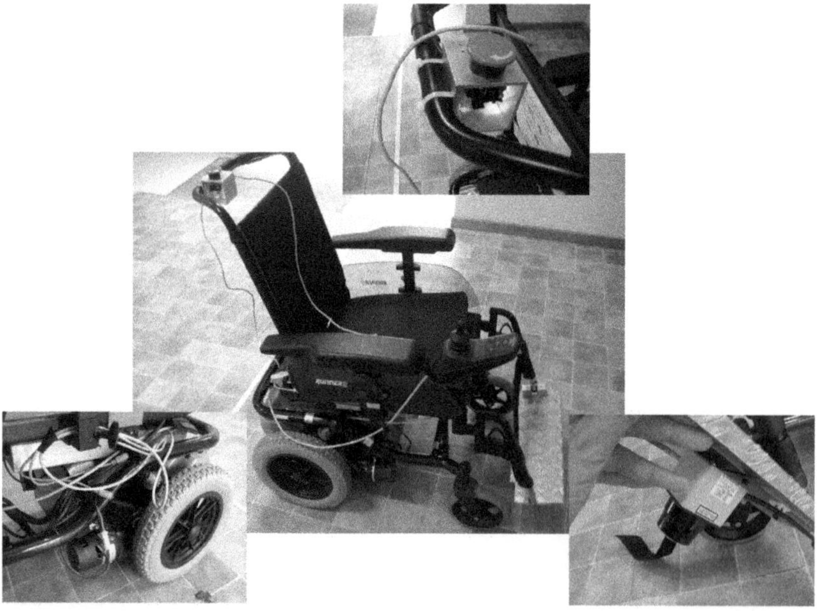

Figure 4
Casa Agevole

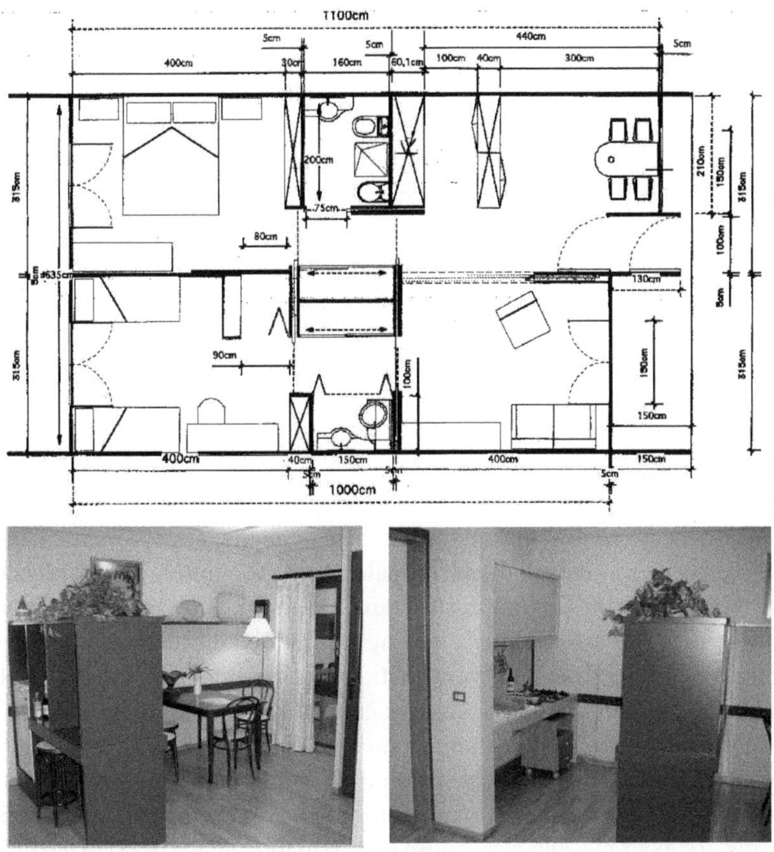

- Higher-level cognitive functions: abstraction, organization and planning
- Mental functions of language: perception of language—spoken language, written language—Expression of language—spoken language
- Behavior and affective status

Also, the exclusion criteria were to present one of among: presence of any behavioral disorders as Delusions, Hallucinations, Agitation/Aggression, Euphoria, Disinhibition, Irritability/Liability, Aberrant motor activity, severe Apathy, severe Depression and Anxiety, and patients clinically diagnosed with global Aphasia.

Volunteers presented different degrees of disability, both physical and cognitive. These disabilities usually ranged from mild to severe, although there was a group of users with minor disabilities to test *how* they accept assistance when their requirements are minor. No user presented visual impairment or tetraplegia, so they all guided the wheelchair with its joystick.

We also tried to include in our target population volunteers whose disability only affected a limited part of their driving skills (e.g., hemiplegia, aphasia . . . etc.) to check how assistance increased or decreased depending on their punctual needs.

With all that information we created functional classes or profiles. For example, the one called Mild impaired is defined for individuals whose MMSE score is between 21 and 24. Memory is ranked as mild or moderate. Their perceptual functions are ranked as attention, higher level cognitive functions, and language (one/two mild or one moderate).

5.1. Facts and Methods

With the given description the typical user defined by the medical team for these experiments is an elderly patient, with cognitive and/or physical disabilities on a mild or moderate degree, co-morbid conditions, functional loss from multiple disabilities, leading to impaired self-dependency. In addition to chronic degenerative conditions such as dementia or post-stroke, standard users might be affected by general declines in motor coordination, visual and auditory acuteness, muscle and bone strength, and sensory perception of stimuli (as heat and cold). Experiments consisted of asking volunteers to perform a given trajectory that was described in qualitative terms, for example, enter the house and go to the kitchen or go to the living room's dining table. There were no visual cues at the house to indicate potential trajectories and no further indication was provided, although there was always at least one team member within reach, as *CARMEN* ultimately yields a panic button to stop all motors in case of danger. All tests consisted of performing one or several trajectories within the house with and without our assistance to check how driving was improved. It needs to be noted that most users had driven a wheelchair before, so the order of the unassisted/assisted runs was not so critical. Final destination in tests was typically not within line of sight, an issue that was important for users with mild to severe cognitive disabilities.

In order to measure performance, we chose a number of punctual and global metrics to evaluate how collaborative control works locally and as a whole (task metrics). We also included some metrics to evaluate acceptance among the target population (psych metrics). The simplest metric concerning navigation is task *success*, meaning that the preset goal has been achieved. Usually, in test courses a maximum time to reach this goal is preset. In some cases, collisions during the course are allowed, if they have been evaluated as *safe*, but counted to evaluate a so-called *degree of success*. Consequently, other popular metrics include task completion time, number of collisions, and distance traveled (Yanco 1998; Parikh et al. 2005). However, in our case collisions were not allowed, so related metrics were dismissed.

All these metrics are global, but we designed some local metrics as well since our collaborative navigation system is mostly reactive. Local metrics rank performance at a given time instant. In our case, we measured driving efficiency (η) in terms of the aforementioned safety (η_{sf}), directness (η_{tl}), and smoothness (η_{sm}) (Urdiales et al. 2009) that loosely correspond to the properties of a navigation function. As commented, they represent how safe a command is, how fast it will get the robot to its target, and how smooth motion will be, respectively. These factors can be calculated locally and also averaged for a whole trajectory.

All commented metrics are related to performance, but it is also necessary to take into account how comfortable the user is during the process. In order to avoid the use of biometrics devices, we chose a number of the so-called psych metrics, typically employed in teleoperation ergonomics. These metrics included Intervention Level, Disagreement, and Joystick Variation. Intervention Level (Pineau and Atrash 2007) is the percentage of time that the user is actually moving the joystick while the wheelchair is moving. In teleoperation, this index is meant to be low, but in collaborative navigation, where the user is meant to participate as much as possible in the process, it should be high instead. Disagreement represents the angle difference between human and wheelchair command vectors. Zero Disagreement means that mobile and human totally agree with each other. A high Disagreement is expected to be related with effort and frustration. In fact, every mobile has a baseline Disagreement that represents how it does not respond immediately and exactly as the driver expects. In our wheelchair, this baseline Disagreement is around 15 percent. Joystick Variation measures changes in the joystick direction larger than a 10 percent and it is reported to be associated with workload (Kondraske et al. 1991).

Finally, since our target group is persons with disabilities, we added a final index: Inconsistency, defined as the variation of the user output when facing similar situations. The goal of this index is to check if a person has a well-defined course of action or, otherwise, tends to be erratic in his/her decisions. A low Inconsistency is expected to be related to users with good cognitive capabilities, whereas a high one is related to random courses. Thus, Inconsistency gives us some idea about the cognitive driving capabilities of the user online and, in combination with Disagreement, provides information about the degree of comfort of the user with the wheelchair motion.

5.2. Results

Results of the presented experiments changed significantly through the commented three years' period, as we progressively improved our system both regarding driving efficiency and user acceptance. Results also present some variations depending on the condition of the patients in each test group. However, as a whole, results during the last year yielded the collaborative control efficiency equal to 76.82 percent for the human efficiency in standalone mode equal to 55.58 percent. The Intervention Level in collaborative mode was equal to 90.78 percent against a 58.82 percent in human standalone mode, proving that the user is notably more active in assisted navigation. Disagreement only grew from 10.41 percent in standalone navigation—what could be considered baseline Disagreement due to the specifics of the vehicle—to 11.68 percent and Joystick Variation also increased slightly from 1.25 percent to 1.74 percent. However, the deviation of these metrics reduced by around a 3–4 percent in both cases. Inconsistency was approximately the same with and without assistance as soon as we introduced the learning-based navigation algorithm to reduce Disagreement, but was not too high: just around a 7–8 percent. It is interesting to note that we neither noticed any major reductions in average trajectory length nor completion time—except in extreme cases or when users could not finish their paths without assistance, since our system does not try to optimize this particular parameter. We noticed, however, that the deviation of both parameters is noticeably lower.

Finally, we performed an ANOVA test to show that improvement was significant. In our last experiment batch—end of the year, when we compared human and collaborative control efficiency, the value of F-test was equal to 9.3699, and the p-value was equal to 0.0022. Thus, we can conclude that user and machine affect each other positively and, hence, cooperate for good.

After tests were conducted, several general conclusions could be extracted from the resulting data:

- Emerging behaviors are consistent with bottom-up reactive control: resulting trajectories tend to be smooth, safe, and less shaky than human control.
- Collaborative control is generally more efficient than human control on an average and even when results are globally similar, there are always locations at the trajectory where the user cannot maneuver alone and would be stuck without assistance even if he could cope well with the rest of the path.
- MMSE is inversely proportional to Disagreement, meaning that persons with less cognitive capacities tend to accept wheelchair assistance better. Obviously, Disagreement is also linked to the time to finish the trajectory: a user with a higher Disagreement takes longer to reach a goal.
- Intervention Level depends on the Barthel Index: persons with better physical capabilities tend to be more cooperative and also present higher smoothness, even though collaborative control tends to equalize performances at these levels (smoothness, directness, and safety).
- If MMSE and Barthel are either very low or very high at the same time, all efficiencies (robot, user, and collaborative control) become significantly lower. This is particularly interesting because it basically means that users with good capacities tend to fight the chair when they realize they are not completely under control. We solved this problem in our last batch of experiments by replacing the PFA navigation algorithm with a learning Case Based Reasoning (CBR) based on one that mimicked the way the user drove as much as possible (Urdiales et al. 2009; Peula et al. 2012).

6. Discussion and Conclusions

It has been reported that elders have a better QoL when they are able to stay in their own homes (Lawton 1990). For elders who remain independent in their homes, the home becomes more than just a place to eat and sleep. Therefore, our goal is to aid the design and development of suitable assistive robotic products and strategies for support of elders to extend their lives in their preferred environment. There is a bigger need to improve the cost-to-benefit ratio of robot-assisted therapy strategies and their effectiveness for many diseases like stroke therapy. Our experimentation with real individuals shows the feasibility and appropriateness of our approach in this area (see §5.2). As stated in §4, robot assisted therapy increases the cost to benefit ratio for many disease therapies like stroke. Our objective was to evaluate possible differences in the recovery and management of resources and mobility in these patients.

One of the most challenging aspects of this whole experimentation phase has been to establish a measure of success as at every step a new aspect or interaction to be considered appeared in the field of robotics. No benchmarks for fully integrated environments exist and the ones benchmarking individual elements are not effective for the scenarios addressed by *SHARE-it*. We had to develop our own approach. In the pilot study that we conducted, we wanted to demonstrate how to effectively use Artificial Intelligence, in special Assistive Technologies, in therapeutics. In this case, using *CARMEN* to complete the rehabilitative intervention providing aid to a conventional system adding features that enable to achieve results not otherwise achievable. The most important outcome is that CARMEN can be used by a population that otherwise would be excluded from operating a power wheelchair on their own. The traditional power wheelchair, in fact, cannot be provided to a patient with some cognitive deficits. The help that the system delivers is customized with *CARME*N for each individual. Our collaborative strategy also allows the patient to guide *CARMEN* through the user's active participation supported by the shared control system that provides only the amount of help needed in different situations, thus ensuring the user's autonomous movement in complete safety.

6.1. Discussion

Robotic interaction should not replace human interaction but rather improve it. This has been already observed by other researchers but when the user has to share some of his/her autonomy with a machine able to make decisions and take actions as it is the case in *SHARE-it*, the user should be able to trust and feel safe using it in his/her daily activities without the participation of skilled individuals supporting the continuity of the services. Taking the appropriate care of older adults is among the major ethical, technological, social, and economic challenges currently faced by ageing populations. The main aim of most proposals made until now is to give the right amount of support allowing elders living in a semi-autonomous way at their places as long as possible. Many of these solutions are based on the assumption of the existence of a good companion (sociable/assistive) robot. An issue to this respect is that there are no long-term studies that allow to confirm or to generalize the results even in the same country.

Good Robots ought to be included in an ecological framework and will need to move beyond task-based interactions and be attractive, affordable, and non-stigmatizing. Accessibility, ease of use, and reliability will also be particularly important for aging users (Forlizzi et al. 2004). Following Jodi Forlizzi et al., the ecology of aging will include people (elders, caregivers, family), products, the built environment including the network, and the community.

A *Good* Robot ought to be connected to its environment—that is, the flow of the information ought to be continuous—it should be adaptive—that is, to the user's need or to changes in the environment—it should be reliable for the user, and the services it provides ought to be dynamic and evolvable. This new generation of *Good* Robots opens a whole set of new ethical questions: Where does the robot's responsibility end? Is it only to send the alarm messages? Who is in charge of the robot? Is it the user or his/her caregiver? Who is legally responsible for the robots autonomous actions?

We firmly believe that the *SHARE-it* approach and architecture will contribute to the new trends among gerontology and/or gerontechnology researchers and other stakeholders that advocate for shift from institutional to preferred environment (home-based) care. In this new scenario, the agent-based technologies and solutions we are proving may positively benefit and help elders, with mind cognitive problems, to age-in-place. In our view the user should only be assisted according to his/her profile: not more, not less (Cortés et al. 2010).

Note

1. http://www.progettarepertutti.org.

References

Aigner, P., and B. J. McCarragher. "Modeling and Constraining Human Interactions in Shared Control Utilizing a Discrete Event Framework." *IEEE Transactions on Systems, Man and Cybernetics—Part A: Systems and Humans* 30, no. 3 (2000): 369–79.

Applica CESEP Alphametrics. "Men and Women with Disabilities in the EU." March 2007.

Branch, L. G., and A. M. Jette. "A Prospective Study of Long-Term Care Institutionalization among the Aged." *American Journal of Public Health* 72 (1992): 1373–79.

Breazeal, C., and B. Scassellati. "Infant-Like Social Interactions between a Robot and a Human Caretaker." *Adaptive Behavior* 8 (2000): 47–72.

Broekens, J., M. Heerink, and H. Rosendal. "Assistive Social Robots in Elderly Care: A Review." *Gerontechnology* 8 (2009): 94–103.

Bruemmer, D. J., D. A. Few, R. L. Boring, J. L. Marble, M. C. Walton, and C. W. Nielsen. "Shared Understanding for Collaborative Control." *IEEE Transactions on Systems, Man, and Cybernetics, Part A* 35, no. 4 (2005): 494–504.

Cauchi, B., S. Goetze, and S. Doclo. "Reduction of Non-Stationary Noise for a Robotic Living Assistant Using Sparse Non-Negative Matrix Factorization." In *Proceedings of the 1st Workshop on Speech and Multimodal Interaction in Assistive Environments, SMIAE'12*, 28–33. Stroudsburg, PA: Association for Computational Linguistics, 2012.

Cesta, A., G. Cortellessa, F. Pecora, and R. Rasconi. "Exploiting Scheduling Techniques to Monitor the Execution of Domestic Activities." *Intelligenza Artificiale (Italian Journal of Artificial Intelligence)* 2 (2005): 74–81.

Connell, J., and P. Viola. "Cooperative Control of a Semi-Autonomous Mobile Robot." In Proc. 1990 IEEE ICRA Conference, Cincinnati OH, 2 (1990): 1118–1121.

Cortés, U., C. Barrué, F. Campana, A. B. Martínez-Velasco, C. Urdiales, R. Annicchiarico, and C. Caltagirone. "Assistive Technologies for the New Generation of Senior Citizens. The SHARE-it Approach." *International Journal of Computers in Healthcare* 1, no. 1 (2010): 35–65.

Espeland, M. P., T. M. Gill, and J. Guralnik. "Designing Clinical Trials of Interventions for Mobility Disability: Results from the Lifestyle Interventions and Independence for Elders Pilot (Life-P) Trial." *Journals of Gerontology Series A: Biological Sciences and Medical Sciences* 62, no. 11 (2007): 1237–43.

FAll detector for The Elder. www.project-fate.eu/, 2012.

Flandorfer, P. "Population Ageing and Socially Assistive Robots for Elderly Persons: The Importance of Sociodemographic Factors for User Acceptance." *International Journal of Population Research* 2012 (2012): 1–13.

Forlizzi, J., C. DiSalvo, and F. Gemperle. "Assistive Robotics and an Ecology of Elders Living Independently in Their Homes." *Human-Computer Interaction* 19, no. 1/2 (2004): 25–59.

Fried, L. P., L. Ferrucci, J. Darer, J. D. Williamson, and G. Anderson. "Untangling the Concepts of Disability, Frailty and Comorbidity: Implications for Improved Targeting and Care." *Journals of Gerontology Series A: Biological Sciences and Medical Sciences* 59 (2004): 255–63.

Fried, L. P., C. M. Tangen, J. Waltson, and A. B. Newman. "Frailty in Older Adults: Evidence for a Phenotype." *Journals of Gerontology Series A: Biological Sciences and Medical Sciences* 56A (2001): 146–57.

Granata, C., and P. Bidaud. "Experimental Analysis of Interactive Behaviors for a Personal Mobile Robot." In *Proceedings of the 15th International Conference on Climbing and Walking Robots and the Support Technologies for Mobile Machines (CLAWAR2012)*, 456–68, 2012.

Guralnik, J. M., A. Z. LaCroix, R. D. Abbot, L. F. Berkman, S. Satterfield, D. A. Evans, and R. B. Wallace. "Maintaining Mobility in Late Life. Demographic Characteristics and Chronic Conditions." *American Journal of Epidemiology* 137, no. 8 (1993): 845–57.

Hagras, H., V. Callaghan, M. Colley, G. Clarke, A. Pounds-Cornish, and H. Duman. "Creating an Ambient-Intelligence Environment Using Embedded Agents." *IEEE Intelligent Systems* 19, no. 6 (2004): 12–20.

Integrated Prevention and Detection Solutions Tailored to the Population and Risk Factors Associated with FALLs (I-DONTFALL). http://www.riscproject.eu/, 2012.

Kamel Boulos, M. N., R. C. Lou, A. Anastasiou, C. D. Nugent, J. Alexanders-son, G. Zimmermann, U. Cortés, and R. Casas, "Connectivity for Healthcare and Well-Being Management: Examples from Six European Projects." *International Journal of Environmental Research and Public Health* 6 (2009): 1947–71.

Kondraske, G. V., G. J. Khoury, K. J. Maxwelle, D. Clarke, and S. Yen. "Telerobotic Network Workstation for System Performance and Operator Workload Monitoring." *NASA JSC, Houston, TX, Tech Rep*, 91-013R, 1991.

Lawton, M. P. "Aging and Performance of Home Tasks." *Human Factors* 32, no. 5 (1990): 527–36.

Mahoney, F. I., and D. W. Barthel. "Functional Evaluation: The Barthel Index." *Maryland State Medical Journal* 14 (1965): 61–65.

Miller, D. *Assistive Robotics: An Overview*, 126–36. Springer-Verlag, New York 1998.

Morris, A., R. Donamukkala, A. Kapuria, A. Steinfeld, J. T. Matthews, J. Dunbar-Jacobs, and S. Thrun. "A Robotic Walker that Provides Guidance." In Proc. 2003 IEEE ICRA Conference, Taipei Taiwan 1 (2003): 25–30.

Parikh, S. P., Grassi, Jr., V., Kumar V. and Okamoto, Jr. J., "Incorporating User Inputs in Motion Planning for a Smart Wheelchair." In Proc. 2004 IEEE ICRA Conference, New Orleans LA, 2 (2004): 2043–48.

———. "Usability Study of a Control Framework for an Intelligent Wheelchair." In Proc. 2005 IEEE ICRA Conference, Barcelona Spain, 4 (2005): 4745–50.

Peula, J. M., C. Urdiales, I. Herrero, M. Fernández-Carmona, and F.S. Hernández. "Case-Based Reasoning Emulation of Persons for Wheelchair Navigation." *Artificial Intelligence in Medicine* 56, no. 2 (2012): 109–21.

Pineau, J., and A. Atrash. "Smartwheeler: A Robotic Wheelchair Test-Bed for Investigating New Models of Human-Robot Interaction." Papers from the 2007 AAAI Spring Symposium, Technical Report SS-07-07, Stanford, California, USA, (2007: 59–64.

Saffotti, A., and M. Broxvall. "Peis Ecologies: Ambient Intelligence Meets Autonomous Robotics." In *Proceedings of the International Conference on Smart Objects and Ambient Intelligence* (*soc-EUSAI*), 275–80, 2005.

Simpson, R. C., S. P. Levine, D. A. Bell, L. A. Jaros, Y. Koren, and Y. Borenstein. "NavChair: An Assistive Wheelchair Navigation System with Automatic Adaptation." *Lecture Notes in Computer Science* 1458 (1998): 235–54.

Supported Human Autonomy for Recovery and Enhancement of Cognitive and Motor Abilities Using Information Technologies. http://www.istshareit.eu, 2007.

Tapus, A., M. J. Mataric, and B. Scassellati. "The Grand Challenges in Socially Assistive Robotics." *IEEE Robotics and Automation Magazine* 14, no. 1 (2007): 1–7.

Tiberio, L., A. Cesta, G. Cortellessa, L. Padua, and A. R. Pellegrino. "Assessing Affective Response of Older Users to a Telepresence Robot Using a Combination of Psychophysiological Measures." In Proc. 2012 IEEE RO-MAN, Paris France, (2012): 833–38.

Urdiales, C., J. M. Peula, M. F. Carmona, R. Annicchiarico, F. Sandoval, and C. Caltagirone. "Adaptive Collaborative Assistance for Wheelchair Driving via CBR Learning." In *IEEE 11th International Conference on Rehabilitation Robotics (ICORR2009), Kyoto, Japan*, 737–42, 2009.

Yanco, H. A. "Wheelesley: A Robotic Wheelchair System: Indoor Navigation and User Interface." In Mittal V. O., Yanco H. A., Aronis J., Simpson R. C. (editors) *Assistive Technology and Artificial Intelligence, Applications in Robotics, User Interfaces and Natural Language Processing*, Springer-Verlag Heidelberg (1998): 235–55.

Unorthodox Thoughts about Good Robots Viewed Pragmatically and Praxiologically

Zbigniew J. Gackowski
Department of Computer Information Systems
College of Business Administration
California State University Stanislaus
Turlock, CA, USA

1. Introduction: Good and Evil versus Quality in General

Contemporary philosophers do not hold scholastic philosophy in high esteem; nevertheless, each epoch contributed something of lasting value. Thomas Aquinas is credited with discovering that there is no thing inherently good or evil, only the *use* of it. Paraphrasing Friedrich Nietzsche (2001), science and engineering are *beyond good and evil.* Good and evil seem to be self-evident by appearance, although only on the surface, but are deceptive in the long run by whatever criteria one uses, despite being the oldest ideas. The devil lurks in the details. We have accumulated much scientific and engineering knowledge (verifiable and replicable in tests), both positive and negative, in managing quality, but the same cannot be said about good or evil, which depends on one's moral principles.

Many of the unquestionably noble attempts to "do good" have frequently turned out to be counter productive, even demoralizing, due to unintended consequences. Praxiology proclaims to be a practical philosophy. For example, one may be even honestly looking for the good security of our children. He declares schools a gun-free zone. In contrast to intentions, a "gun-free zone" sign is a public invitation to a deranged individual dreaming about the fame from killing more than were killed in a Norway shooting, to play God with children's lives with impunity. He might have had second thoughts about invading a place where he knew the occupants could not defend themselves. Such situations are the fate of good intentions of some voters and politicians gone wrong, but well-trained and experienced individuals, such as the Harris County, Texas, sheriff, chimed in, "We need funding and well-trained police officers. My campus, fortunately, had funding and a well-trained police force on site but still witnessed a perfect fiasco. A youngster stole a gun from a well-trained retired police officer, penetrated two campuses, and shot at my daughter in the hallway. If I could carry a gun, he would not have had a chance, but he escaped unharmed, apprehended only a long time later after the damage was done. This is not a conceptual schema to enliven an academic paper (as some editors advise). In philosophical terms, it is existence defined as reality perceived in experience." Humans are extremely gullible about trusting in short-lived solutions by doing "something" instead of doing "right."

Robots are controlled by software, but how gullible humans might be attested to by innumerable learned discussions that were conducted by professionals and lay people (at the onset of computing) about whether software should be free or whether it should be the subject of intellectual property rights. Then, most respondents supported free distribution of software for the "common good." If implemented, it would kill the goose that lays the golden eggs: the giant and highly profitable software industry, Microsoft and the like, that now practically drives the development of computer hardware. However, the situation is more complex. On the other hand, we witness the successful embedding of free Unix derivatives in commercial products. Most modern televisions, virtually all smart phones, tablets, routers—both home and industrial—are based on Linux, FreeBSD, or other such OSs. Yes, the profit motive pushed the industry greatly in its infancy, but free software also clearly has established itself in infrastructure applications. What is the ultimate moral? In economy, the freedom of contract is the best solution, not "well-intentioned" but "ill-conceived" regulations by those who claim they know better. Do-gooders tend to slow down progress.

For the good, for equality, or for fraternity, much blood was spilled but never attained. Few know that perfect equality of all states of matter in nature would result in thermal death of the universe and would halt human progress. Similarly, the biblical Cain–and–Abel story did not forebode a bright future for fraternity. Nature does not know equality; it evolves by mutations and diversity. Finally, what do we take away from this discussion? It is this: "Good" is an ambiguous term to which each of us may ascribe a different definition. We need more precision.

Thus, the term "good robot" may serve as a shortcut for the well-designed robot, in contrast to that which is poorly designed: those that better "connect perception to action" (Brady 1984). But this is insufficiently precise. We need, instead, high-quality robots that meet the fundamental triad of praxiological criteria of assessment: ethics (the extent to which it meets the applied ethical standards), effectiveness (the extent to which it serves the purpose of action), and efficiency (the ratio of the results with regard to the purpose and the measure of resources engaged and/or consumed). Quality, at least quality of operations, is better defined than goodness. This article posits that these praxiological criteria offer an improved framework for research in robot quality (RQ) or in well-designed robots rather than merely good robots. However, the interpretations of those criteria are far from established.

2. The Three Es: The Praxiological Triad

Tens of publications on this subject exist. For the sake of brevity, only the first and the last that were found by Google Scholar are quoted:

- Gasparski, Ryan, and Kwiatkowski (2011), in his editorial, writes, "*. . . the praxiological and ethical triad: the three Es or effectiveness, efficiency, and ethicality* (p. 9), where *ethicality . . . constitutes an irremovable axiological context of the other two*" (p. 34), in contrast to the wording of the title of Volume 17, which suggests ethics as an add-on to praxiological dimensions, where ethics was never or rarely absent.
- Ryan (1996), in his introduction to Volume 5 writes, "*It is the explication of the 'triple E': Efficiency, Effectiveness, and Ethics to which this Volume is dedicated*" (xiii), and Gasparski and Ryan (1996), in his editorial, quotes Józef Maria Bocheński, a philosopher and Dominican friar: "*. . . actions are performed . . . against the 'double E criterion' composed of: (i) effectiveness . . . and (ii) efficiency. . . . The seriousness also requires that the social axiological context . . . is taken into account. . . . Here, the appraisal of actions depends on the 'third E' for ethical (moral) values . . .*" (vii–viii).

These statements, including those by Bocheński, put ethics at the end of the triad, and they differ with regard to the sequence of the other criteria.

Philosophers are remote from a uniform interpretation of the praxiological criteria. This paper posits that, within the context of human-controlled operations, or human actions, the triad of praxiological criteria needs be revised to ethics, effectiveness, and efficiency (Gackowski 2012). This is supported by the following rationale:

- Any human-controlled operation is de facto an ethical or moral statement by the one who triggered it, as being a right or justified action under the given circumstances. It may be subjective, judged differently by others, but, nonetheless, it is the strongest statement a human can make; not in thoughts, utterances, or words, but in action.
- Many actions are deliberately triggered for ethical reasons; then ethical considerations come first.
- Once action is considered to be justified (rightly or wrongly), effectiveness rather than efficiency with regard to the purpose should be considered. If its attainment is not acceptable plausibly, it makes no sense even to think about efficiency of action.
- Considerations of efficiency apply to premeditated actions, when humans may act regardless of efficiency (the amount of resources engaged and consumed) out of sheer rage, heroic effort, and even in suicidal terror attack, where the attained psychological satisfaction outweighs anything else: liberty, possession, or the actor's life.
- In organized, hence premeditated actions, at least it is assumed, rightly or wrongly, that benefits will outweigh the cost of the engaged and consumed resources.

Thus, the three Es, the triad of praxiological criteria of assessments of human-controlled operations, provide an axiological context where only the first two are irremovable, while all three are rigorously hierarchically interdependent, and their hierarchy is not subject to liberal interpretations. The term "human-controlled operations" is used on purpose; however, robots may differ vastly with regard to the level of human control. Robots may be controlled directly and entirely by humans, such as robots used to disarm booby-traps, the space robot Curiosity, or intelligent robots that act autonomously; nevertheless, even then, they are controlled indirectly by human design. The future, however, is not clear in this respect. What might happen when intelligent autonomously working robots will be allowed to design other intelligent autonomously working robots? This is only a matter of time. When it comes to intelligence, (Ray Kurzweil's 2012) work about a kind of replication of human mind has already begun in earnest, particularly that Kurzweil recently joined Google with its vast resources. Now, we should reflect about laws of robots or robotics.

3. Laws of Robotics

Some so-called laws of robotics have been articulated, which, however, are not quite laws of nature, as laws that neither can be imposed nor enforced by humans; frequently, they represent some wishful thinking. Below are some of the best-known laws of robotics (Laws of Robotics 2013), with my unorthodox comments.

In 2011, the Engineering and Physical Sciences Research Council (EPSRC) and the Arts and Humanities Research Council (AHRC) of Great Britain jointly published a set of five ethical "principles for designers, builders and users of robots" in the real world, along with seven "high-level messages" that were intended to be conveyed (not listed here), based on a September 2010 research workshop:

1. Robots should not be designed solely or primarily to kill or harm humans. (My comment: *A hopeless postulate; an example of wishful thinking with no chance of implementation. The currently famous, or infamous, drones are examples of an entire war industry working effectively in the opposite direction, and no one will be able to stop it. Such drones are only the beginning. What about insect-size and insect-like mini or micro drones that swarm over the target? They have been designed to be heavily attracted to any living tissue. They sting and paralyze or incapacitate the target in other ways temporarily or forever. No controlling and enforcing authority exists to prevent it from happening. The harsh laws of war require counter measures with likely even more vicious robots.*)
2. Humans, not robots, are responsible agents. Robots are tools designed to achieve human goals. (Author's comment: *No controlling and enforcing legal authority exists to prevent the development of disobedient robots. Let us, however, assume that some obedience is built into the robot by design; now we face another question: obedient to whom? (Even to the most irresponsible among human beings?) This principle implies two things: Limitations on the intelligence of the robot, and limitation on being obedient to specific humans only. Again, if one continues this train of thought, one may question who they would be obedient to and by what criteria? In social reality, it means to someone of special trust, to the so-called elite. Of course, the elite would be those in power or those whom some power has been delegated. This leads us to the governing elite and, finally, to the commander-in-chief, as it is with the contemporary drones. Finally, this may lead to an ultimate dictatorship, where obedient robots will subdue any manifestation of dissent—a new version of "The Year*

2084." Humans have a proven record of difficulty thinking a situation through to its concluding end.)

3. Robots should be designed in ways that assure their safety and security. (Author's comment: *We may try this with similar imperfection as we do it for ourselves.*)
4. Robots are artifacts; they should not be designed to exploit (*neither politicians*) vulnerable users by evoking an emotional response or dependency.
5. It should always be possible to tell a robot from a human. (Author's comment: *A classic unenforceable wishful thinking as long as mankind does not open the Pandora's box of intelligent fully autonomous robots reproducing their likes.*)

Mark W. Tilden, a notable robotics physicist who was a pioneer in developing simple robotics, developed three guiding principles/rules for robots:

1. A robot must protect its existence at all costs. (*An example of how modest professionals are.*) (Author's comment: *A down-to-earth statement; the future will follow this path; however, a warning is necessary here: Intelligent and fully autonomous robots inevitably must start competing with humans for resources.*)
2. A robot must obtain and maintain access to its own power source. (Author's comment: *This is obvious, and no one will be able to prevent it short of a war of human's contra-intelligent and fully autonomous robots that were designed this way.*)
3. A robot must continually search for better power sources. (Author's comment: *Intelligent, fully autonomously functioning robots will pursue it on their own, which leads to competition with humans.*)

What is notable in these three rules is that they are basically rules for "wild" life, so, in essence, what Tilden stated is that, what he wanted was "proctoring a silicon species into sentience, but with full control over the specs." (Author's comment: *Again, this is the unknown future once the Pandora's box of intelligent autonomous robots reproducing themselves is opened.*) Not plant. Not animal. Not human. It is something else (see http://en.wikipedia.org/wiki/Tilden%27s_Laws_of_Robotics).

The best-known set of laws is Isaac Asimov's *Three Laws of Robotics*. These were introduced in his 1942 short story, *Runaround*, although they were foreshadowed in a few earlier stories. (These laws came first chronologically; however, as fiction, they pertain to fully intelligent robots, which, in his time, did not exist, but they are becoming reality now. They assume that robots are servants of and are subservient to humans, even though they are of comparable intelligence. Kurzweil raises an ethical issue: Once we have intelligent machines that are conscious at our level, how can we state that we humans are superior to them? The argument is that, if you build a machine that exactly simulates your brain, you have created a duplicate of yourself. Is it not a conscious being comparable to a biological human? Would it be ethical to turn it off?)

The Three Laws are as follows:

1. A robot may not injure a human being or, through inaction, allow a human being to come to harm. (Author's comment: *Wishful thinking, as the First Law by EPSRC.*)
2. A robot must obey the orders given to it by human beings, except where such orders would conflict with the First Law. (Author's comments: *Hopeless, wishful thinking, as the Second Law by EPSRC.*)
3. A robot must protect its own existence as long as such protection does not conflict with the First or Second Laws. (Author's comment: *Certainly, this is inapplicable to military killer robots, and who will prevent them from being developed?*)

In later books, a zeroth law was introduced: 0. A robot may not harm humanity, or, by inaction, allow humanity to come to harm. (Author's comment: *All the previous objections fully apply. Here, however, philosophers beware. This definition defies our current level of understanding of the philosophy of science. It is a similar failure as Immanuel Kant's second categorical moral imperative, which, by current standards, is an intellectual construct that, from its inception, is faulty by design; it refers to about seven billion nebula of points of reference—the entire humanity in all its variety, which make reliable observations and cohesive statements impossible because anything can be proven.*)

4. Quality of Operations

Quality in general is a multidimensional space of specific qualities, meant as specific attributes or dimensions (within the quality space) of defined granularity that describe something under assessment with regard to its required intrinsic aspects, its use, and other expectations that are subject to quality assessment by users, handlers, and other stakeholders. Some of the qualities are characteristics (necessary distinguishing qualities); they may be categorical, situation-specific necessary, desirable, or only optional. With robots, it pertains to their capabilities with regard to perception of reality, execution of action, and intended connection of the perception to action. Robots' hardware (or embodiment) entails all sensors, the computer, the execution mechanism, and, for example, the gigantic communication hardware needed to communicate with the robot Curiosity in space.

The *state of quality of a robot* may be defined by a vector of properties (states) of its intrinsic qualities and use and handling requirements. Each quality can take on one out of two or more distinguishable properties that may be Boolean, enumerated, or ordered sets of numbers that rank properties of gradable qualities. The *state* (not measure) *of quality of operations* performed by a robot may be defined by a set of vectors of significant properties of significant qualities that are intrinsic to significant factors or parameters of its use with regard to the purpose of its operations. One desires optimal or acceptable results of operations with regard to ethics, effectiveness, and efficiency. One can assess the state of quality of operations only indirectly by the impact of its aspects on results of operations, not by properties, as practiced by Total Quality Management (TQM). In operations, significant properties of significant qualities of significant factors are at their *optimal or acceptable level* when the results are optimal or acceptable, which determines the optimal or acceptable states of factors' properties; otherwise, they are of no merit on their own.

5. Praxiological Assessment of Operations

Praxiology is a general theory of human conduct (Gasparski 1988), which is assessed by a triad of criteria: ethics, effectiveness, and efficiency of action.

- ***Ethics*** (sometimes reduced to legality, only the minimum of morality), for those who seek and wish to follow ethical guidance, suggests which operations are worth conducting, their purposes, means, modes of execution, and the lower levels of their assessment criteria. For instance, is it acceptable that an information system reveals names and addresses of gun owners (then beware), or the names and addresses of the impaired and unarmed (easy marks for the bad guys)? While legality may be enforced, ethical guidance cannot be mandatory. Decisions may be voluntarily guided by agreed-upon ethical standards. Some of them may be rational, such as reciprocity, by Confucius, or the rational categorical imperative, by Immanuel Kant, others may be based on cultural norms or religious beliefs. This paper refers only to rational ethics, as the only thing of likely universal appeal. Religious beliefs are also considered, but they are not of universal appeal, particularly when theological teachings of one of the largest denominations say that faith is a gift of grace.

- ***Effectiveness*** measures the extent to which a factor significantly contributes to the purpose of operations or actions (e.g., *profits, victory* [defined by unconditional surrender], *or a mission*). With ethics bypassed, operations may be planned and conducted with regard to effectiveness only; with an all-out effort, regardless of other consequences, they may be planned and conducted as heroic acts of determination in special operations or suicidal terror acts, or they may be rooted only in desperation and emotions. The organizational purpose is subject to the *postulate of teleological perspectivism and restricted relativity of assessments* that pervades and rules supreme over all human-controlled operations, including research for extension of knowledge (Gackowski 2012, 107) and all communications that serve operations. Factors are subject to assessment within
  - the teleological ***perspectivism***, as defined by Nietzsche (2007), and
  - the ***context***—the frame of reference (in physics) within which each factor or aspect of quality is used in real situations and assessed within Special Relativity, as defined by Albert Einstein (1961). This suffices in the earthly environment. Further exploration of space (such as, with the robot Curiosity but beyond our solar system, with international competition and/or collaboration) may soon require a theory of operations that accounts for the general theory of relativity. There, *time and space lose their traditional role as fixed frames of reference to become reduced to structural qualities of the field* (Einstein 1961) under investigation. Changes are viewed and assessed within such a context. Measurements depend on the selected perspective, point of reference, frame of reference, and the chosen yardstick. Entities and their qualities are assessed equally unless the purpose and frame of reference change. In human-controlled operations, purpose serves as the main point of reference; if purpose and circumstances change, even physically identical entities are assessed differently by observers, decision makers, and actors.
- ***Efficiency*** measures whether the output is worth more than the input and whether it is conducted economically with regard to the engaged resources, incurred cost, outlays, or expenditures at an acceptable level of the ratio of benefits to inputs.

According to the current state of knowledge, robotic operations should be designed by

- employing tried and verified methods and rules of proceeding, with decisions made by rational and rule-following choices, as defined by James March (1994), with bounded rationality,

- as defined by H. A. Simon (1956) with regard to limitations of cognition, and
- as proven by Tadeusz Kotarbiński (1961, 189–201) from an opposite perspective. Perfect rationality is unattainable. If rigorously pursued, it leads to a logical paradox that precludes any action. It is referred to as the Kotarbiński-Simon problem of unattainable perfect rationality (Tsing 1993)

- preventing irrational actions that can be potentially harmful
- ethical, if possible, or automatic arbitrage in making a decision without crippling delays. (It may be impossible to make an unambiguous choice between near-simultaneous events under a deadline (Denning 2007) because systems may hang up).

6. Categories of Users, Stakeholders, and Perceptions

At least four categories of users differ starkly by how they assess quality:

1. ***Organizational users***; their executives (should) focus on ethics, effectiveness, and efficiency of operations. Those for profit, if effective, must be at least sufficiently efficient to generate profit that does not apply to those not for profit. The latter do not have a natural threshold benchmark (e.g., *loss of profitability leading to bankruptcy*) for the efficiency of their operations; however, the most donating users will take into account their efficiency when considering further support. When efficiency is the sole criterion, there is an obvious danger that the ends may justify the means, even mass murders, as history abundantly attests.
2. ***Individual users within organizations*** focus mainly on their comfort as users.
3. ***Individual consumers*** of a robot, or of the service it provides, interpret their entire experience at all the touch points related to its use and handling.
4. The ***environmental stakeholders*** may not be consumers at all but are affected by different aspects of a robot's operations to such a degree that they care about it.

Reality appears differently when perceived directly or indirectly. From the top, qualities viewed along the chain of command appear as a beautiful bell-curve distribution of probabilities, like the almost-perfect sphere of the planet Earth when viewed from a spacecraft. The lower one descends, the more rugged the landscape becomes—less abstract and more specific. When one reaches the bottom (such as the submicroscopic level of matter), smoothness and continuity disappear completely, and reality becomes discrete; probabilities are replaced by binary choices: whether to act or not and whether to be or not to be, with life at stake.

At the operational or execution level (patient, surgeon, soldier in special operations, or a policeman), most qualities are perceived dichotomously (interpreted, misinterpreted; relevant, irrelevant; significant, insignificant; accessible, inaccessible; complete, incomplete; true, false; and so forth), where each choice immediately impacts the behavior of the actor, including a robot. Those directly engaged do not ignore probabilities but prefer not to rely on them; they aim at certainty. This contrasts starkly with the perspective of high-level administrators, politicians, even generals with too much time on their hands. Dichotomously viewed properties (states of qualities) are not stored as data values in relational databases where they are associated only with probabilities of accuracy, rather than as they are perceived, acted upon, and considered at the execution level.

7. Use Requirements and Materiality of Robotic Operations

The optimal or acceptable results of operations determine the optimal or acceptable states of factors' properties, which, on their own, are of no merit. For instance, the sharpness of a knife is a desirable property if the purpose is to cut easily, but it is a risk factor in the hands of residents who need assisted care. Only some qualities are intrinsic to factors; most are situational and situated use and handling requirements (a distinction rarely made). They may differ substantially when moving from one field of expertise to another. The way entities are designed, used, or handled; they might, in practice, make a significant material difference; it pertains to entire robots as to states (properties) of their quality aspects.

Only *operational materiality* adequately measures the importance of any factor. It is determined by the difference in the measure of results, when operations are conducted with and without a factor, including a robot (Gackowski 2012). Principles of Auditing (Whittington and Pany 2008) provide examples of materiality. It ought to be *significant*—exceed some threshold of significance as defined by policy. *Significant operational materiality* is not only primary categorically or universally necessary but also the fundamental, central, and most pervasive requirement related to use of factors. Materiality reigns supreme over all the other quality aspects (except for research of unknown phenomena). For practitioners, anything of insignificant operational materiality does not deserve much attention, in contrast to significant, as defined above. Materiality is *fundamental* because it is the only requirement that provides a sufficient reason for a factor to be considered; it is *central* because it also ranks or orders partially (asymmetric, transitive) all factors for risk assessment; and it is the *most pervasive* use requirement because it lends itself (materiality) to each necessary quality requirement of the same factor and to each necessary task-specific companion factor.

Analysis of factors (also robots) and related use requirements asks the following: (a) Which of the qualities affects operations directly or only indirectly? (b) Which are necessary or only desirable? (c) Which of the necessary are primary or secondary? (d) Which of the primary are universally necessary or only situation-specific necessary? (e) Which of them affect operations qualitatively or mainly quantitatively? (f) In what order should they be examined? (g) What is the extent of their impact-materiality? Most studies never ask these questions; while depending on the answers, the same quality needs to be treated differently.

Directly means that changes of use requirements directly affect the decision situation, the way decisions are implemented, and/or the results of operations. *Indirectly* means that such changes contribute to the state of quality of their respective higher-order factors (ultimately the direct ones).

Necessary use requirements are mandatory. If any of them cannot be met, it precludes not only its use but also further examination of the affected factor. Such a binary distinction is rarely made in current studies, and their logical interdependency is practically ignored.

Primary use requirements are those necessary use requirements that are determined by the nature of the situation independently of the will of the decision maker, such as the ones enumerated below: the necessary or sufficient ones. *Secondary use requirements* depend on the chief executive decision maker; they usually pertain to ethics and efficiency of operations.

Necessary requirements are of the highest importance and of the same ultimate consequences; if not met, the affected factor is not usable for operations. Some are always necessary; hence, they are

- universally necessary or categorical, while the remaining ones are
- other situation-specific necessary requirements. Only the categorical ones can be enumerated.

What makes specific robots effective, usable, and useful in operations at the execution level (all are dichotomous)? They are

1. *recognizable* (by device, operator, or actor)
2. *relevant* (considered by the mental/formal model of operations)
3. *meaningful* (make a pragmatic difference in the results of operations)
4. *significantly material* (exceed the threshold of significance); only then is a robot *pragmatically effective*
5. *on-site available* (where needed for action)
6. *on-time available* (when needed before an opportunity of action expires)
7. *reliable* (whether it works and one may rely on it)
8. *effectively usable* (whatever meets all the above and other necessary situation-specific requirements—a sufficient reason for effectiveness)
9. *effectively useful* (whatever is effectively usable and engaged in operations—a sufficient reason for effective usefulness; otherwise, it is an *effectively useful robot in waiting*).

Changes of significant qualities may require a partial redefinition of the decision situation, for example, *adding or deleting column(s)* or *row(s)* from the decision-situation matrix. These are *qualitative changes*. Changes of states of gradable qualities affect *mainly quantitatively* the results of operations, the decision situations, and/or the way decisions are implemented.

Mainly means here that gradable use requirements that quantitatively affect operations do not exceed their respective acceptable limits (e.g., the blood alcohol *limit that defines driving under the influence*) or reach other critical states (e.g., *melting, evaporating,* or *freezing* in deep-water-well capping); they may trigger qualitative changes (e.g., *an explosion*).

8. Summary

The pragmatic and praxiological view of the quality of robots suggests the need for the following:

1. Substitution of the term "good" by the term "quality" as a more developed concept.
2. Partial revision of the interpretation of the three Es in the praxiological criteria of assessments: ethics, effectiveness, and efficiency as a hierarchical triad.
3. The distinction of at least four categories of users and stakeholders that differ starkly by how they assess quality:
    a. Organizational users who are focused on ethics, effectiveness, and efficiency of operations.
    b. Individual users within organizations who are focused mainly on their current comfort as users.

   c. Individual users as independent consumers of products or services who interpret their entire experience at all the touch points related to the use and handling of information.
   d. The environmental stakeholders who may not be consumers at all but are affected by information quality management to such a degree that they care about it.

4. High-quality robots or good robots that fit defined ethics, effectiveness (with regard to the organizational purpose), and efficiency of individual use.
5. The realization of how differently quality is perceived along the chain of command, from a bell-curved distribution of probabilities at the top to the discrete dichotomous choices at the execution level.
6. The focus on operational materiality that adequately measures the importance of a robot and any aspect of its quality, as well. Materiality ought to be significant. It is a fundamental, central, and most pervasive requirement that is related to use of factors; it reigns supreme over all the other quality aspects (except for research of unknown phenomena).
7. An operational distinction of what makes a robot effectively usable and useful in operations. All are dichotomous. Robots must be recognizable (by the operator), relevant (to the model of operations), pragmatically meaningful, significantly material, on-site available, on-time available, reliable, effectively usable, and effectively useful (as assessed by the stakeholders) if engaged in operations.

Thus, we wish for good robots, while we need high-quality robots, but we may also get intelligent and evil robots. Worse, there is no, nor is there likely to be, a controlling and enforcing legal authority to prevent it from happening. Of course, it is up to the reader, the researchers, to pass the final judgment on the presented suggestions; however, they should be discussed. This paper posits that there is merit in considering the revised praxiological perspective in viewing all actions, not robots only.

9. Acknowledgment

Rick Mayforth, an avid reviewer of publications about technology, should be fully credited for bringing up the need for taking a critical position with regard to the many articulated laws of robotics.

References

Brady, M. *Artificial Intelligence and Robotics*. Artificial Intelligence Laboratory, MIT, A.I. Memo 756, 1984.

Denning, P. J. "The Choice Uncertainty Principle." *Communications of the ACM* 50, no. 11 (2007): 9–14.

Einstein, A. *Relativity—The Special and the General Theory.* New York: Crown Publishers, 1961.
Gackowski, Z. J. *Informing for Operations: Framework, Model, and the First Principles.* Santa Rosa, CA: Informing Science Press, 2012.
Gasparski, W. "Praxiology." In *System and Control Encyclopedia*, ed. M. G. Singh. Oxford: Pergamon Press, (1988): 3860–65.
Gasparski, W., and L. Ryan. *Human Action in Business: Praxiological and Ethical Dimensions.* New Brunswick, NJ: Transaction Publishers, 1996.
Gasparski, W., L. Ryan, and S. Kwiatkowski, eds. *Entrepreneurship: Values and Responsibility*. New Brunswick, NJ: Transaction Publishers, 2011.
Kant, I. *Encyclopædia Britannica, In Encyclopædia Britannica Online Academic Edition*, 2012. Retrieved from http://www.britannica.com/EBchecked/topic/311398/Immanuel-Kant.
Kotarbiński, T. The Property of a Good Plan. *Methods* 13, no. 51–52 (1961): 189–201.
Laws of Robotics. Wikipedia, 2013. Retrieved from http://en.wikipedia.org/wiki/Laws_of_robotics.
March, J. G. *A Primer on Decision-Making—How Decisions Happen.* New York: Free Press, 1994.
Nietzsche, F. *Beyond Good and Evil: Prelude to a Philosophy of the Future*. New York: Cambridge University Press, 2001.
———. *Encyclopedia Britannica, Encyclopedia Britannica Online Academic Edition,* 2007. Retrieved from http://www.britannica.com/EBchecked/topic/311398/Immanuel-Kant.
Simon, H. A. "Rational Choice and the Structure of the Environment." *Psychological Review* 63, no. 2 (1956): 129–38.
Tsing, Z. "Philosophy of Technology: Epistemological or Praxiological? Some Lessons from Chinese Philosophy of Technology." In *Praxiology: The International Annual of Practical Philosophy and Methodology*, vol. 2, ed. T. Airaksinen and W. Gasparski, 243–55. New Brunswick, NJ: Transaction Publishers, 1993.
Whittington, R. O., and K. Pany. *Principles of Auditing & Other Assurance Services*, 16th ed. New York: McGraw-Hill Irwin, 2008.

Part Two

Good Robots—A Technological Perspective

Towards Application-Oriented Navigation

Adam Borkowski
Polish Academy of Sciences (IPPT PAN)
Warsaw, Poland

1. Introduction

The history of intelligent mobile robots can be traced back to the 1950s. Their predecessors were artificial turtles, developed within the framework of cybernetics—the conglomerate of control theory, information science and biology—pioneered by Norbert Wiener in the late 1940s. Perhaps the best known artificial turtle is the Tortoise created by William Grey Walter (Walter 1950). It was a simple wheel-based robot that moved toward a weak light and ran away from a bright light. Despite primitive sensors—a photocell and a contact gauge—the Tortoise was able to avoid obstacles and to simulate insect-like behavior. Further coverage of the history of intelligent mobile robots can be found in the lecture notes by Lynne Parker (2012).

The origin of mobile robotics (MR) in Poland is related to the research project CPBP.02.13 entitled *Systems with Artificial Intelligence for Working Machines and Vehicles*, launched by the State Committee for Scientific Research (KBN) in the late 1980s. Within the frame of this project, the first Polish intelligent mobile robot Ulisses (Figure 1) was built by Andrzej Wołczowski at the ICT-PWr[1] in Wrocław (Wołczowski 1992).

A tiny ZX Spectrum served as a board computer for the robot. The computer had the 8-bit processor Zilog Z80A running at 3.54 MHz.

Figure 1
Robot Ulisses (courtesy of A. Wołczowski)

The unit was equipped with a 16 KB ROM and was connected by the RS-232C link to a desktop computer. The Ulisses was endowed with short-range infrared detectors and contact gauges, enabling it to detect obstacles at the range of 0–1 meters. The software, developed at the PSA–IPPT PAN[2] in Warsaw, enabled the Ulisses to explore its environment and to plan and execute simple trajectories.

The progress in the MR-domain can be appreciated by comparing the Ulisses to contemporary robots. An example shown in Figure 2 is the RoMEG^{AT}, the universal indoor-outdoor robot, built by Rafał Chojecki of the MEG-company[3] in Warsaw, according to the specification given by the author. The robot is mounted on a robust six-wheeled chassis. It can surmount slopes inclined at up to 15 degrees and it can negotiate sills reaching 10 cm in height. Its on-board computer ARK–1388 runs on a 1.06 GHz Core 2 Duo CPU and it has a 2 GB RAM. The sensors mounted on the RoMEG^{AT} include the Sick laser scanner LMS 111 (Sick 2012), the Kinect TOF-camera (Kinect 2012), the IMU-module MTi of Xsens Technologies (MTi 2012), and a GPS-receiver and rotary encoders mounted on the driving wheels. The robot communicates over WLAN with the personal computer, where the higher level control and the GUI

Figure 2
Robot RoMEG^{AT}

are deployed within the framework of the ROS operating system (ROS 2012). The RoMEG^{AT} monitors a range of 0–5 meters, is able to build a map of unknown terrain, to localize itself on the map, to plan collision-free paths toward given goal positions and to execute such trajectories at a speed of up to 2 m/s.

The RoMEG^{AT} is a mobile robot typically used for research purposes today. It is well-suited for developing new procedures in the navigation and exploration of environment domains. On the other hand, it does not have any functions that can be applied to a real-life situation. Till today, few mobile robots broke the barrier between research labs and applications. The most popular one in the market is probably the vacuum cleaner Roomba (2013), a self-propelled circular unit that autonomously finds its way among the pieces of furniture in a room. Other applications include lawn mowers (Mower 2012), security robots (Vigilant 2012) and robotized toys (RoboToys 2012). Impressive progress has been achieved in remote-operated robotics, excellent examples of which are the Mars rovers (e.g., Curiosity 2012) and military drones (e.g., Predator 2012).

However, it seems that mobile robot technology has reached the advanced stage of development that will enable a much broader range of applications than has been possible in the past. In order to achieve such breakthroughs, the MR community should reconsider the emphasis given to particular topics in research. It appears that at present, pure cognitive problems have taken precedence over application-oriented issues. In chapter 2, we intend to prove this hypothesis and to propose some remedies. Further discussion of the present state of MR-research can be found in Ambroszkiewicz (2010).

2. Mapping

A description of the environment in Mobile Robotics is customarily called a map, and the process of describing the exploration of the unknown environment is called mapping. Maps are of two types: metric maps and topological maps. The former can be either vector-based or raster-based, whereas the latter are usually graph-based. Nowadays, there is a tendency for complex information to be stored in a map. Thus, an object-oriented map allows the robot to store multiple attributes that describe particular elements of the environment. It has been shown in Faderewski (2010) that this paradigm is very useful in urban areas. Siemiątkowska et al. (2011) describe a hybrid map that can be seen as a union of grid-based and object-oriented maps.

Till now, the main attention in MR was paid to mapping. This is understandable, since acquiring a map from scratch is cognitively more demanding than a mobile robot controlled with the help of a given map. It was even shown that a map of unknown terrain can be built at the same time as self-localization. Such an incremental procedure, called Simultaneous Localization and Mapping (SLAM),[4] is described in Thrun et al. (2005).

Constructing a map without any prior knowledge of the environment is time-consuming; it distracts the robot from service tasks and diminishes its efficiency. This seems to be the first argument in favor of replacing mapping with maintenance of map (see Section 2.3). The second argument is that nowadays digital maps are easily available. Let us consider these issues separately for two classes of mobile robots: the robots working indoors and the robots working outdoors.

2.1. Indoor Mapping

It is clear that all buildings erected during the last four decade have been designed with the aid of CAD software. Hence, their documentation is available in digital form. Moreover, the Architecture-Engineering-

Construction (AEC) industry has agreed recently on a standard that unifies the way in which each building is described. This standard is called Building Information Model (BIM). Among several books that describe details of BIM and that give examples of its applicability, the monograph by Eastman et al. (2008) is the most comprehensive one.

The BIM is constructed according to an object-oriented parametric description. Components of buildings are presented as hierarchically nested objects that have attributes and application rules. Thus, a particular building consists of instances of object classes, such as walls, floors, and so forth. Figure 3 shows four hierarchically nested layers of the BIM. The top layer gives access to specific sub-domains of knowledge related to the building. Until now, the most frequently used were the domains of architecture, structural analysis, heating, and so forth relevant for the AEC industry. It is reasonable to expect that for intelligent buildings of the future the domains of building control and telecommunication will become increasingly important.

The BIM is described in a data modeling language called EXPRESS (Schenck and Wilson 1994). All objects stay in hierarchical parent-child trees, where children inherit properties from parents. The model is open in the sense that the user is allowed to introduce one's own objects, attributes, and relations. This makes the adaptation of BIM for MR purposes easier.

The most relevant in this respect is the BIM class called Space. It describes a part of the building, bounded physically or virtually, that serves a certain purpose. Such a purpose can be stored in the attributes of Space, like Name or Description. Reading such attributes, the mobile robot becomes aware of the functions carried out by particular parts of the building. In order to understand the topology of building, the robot can take advantage of the relation Connects present in BIM data. This

Figure 3
Main layers of the BIM–model

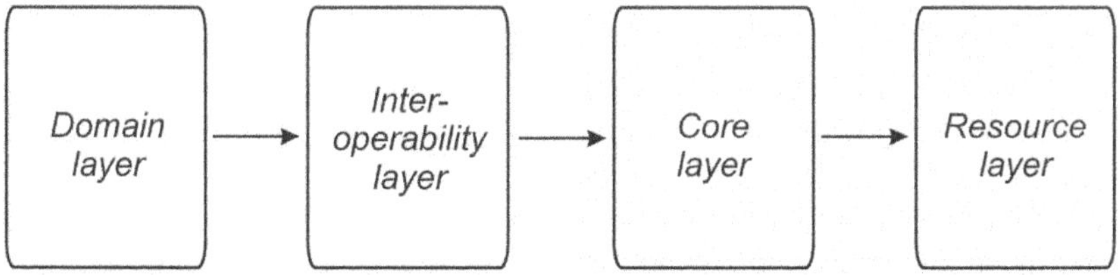

Figure 4
Data transfer from AEC–format to MR–format

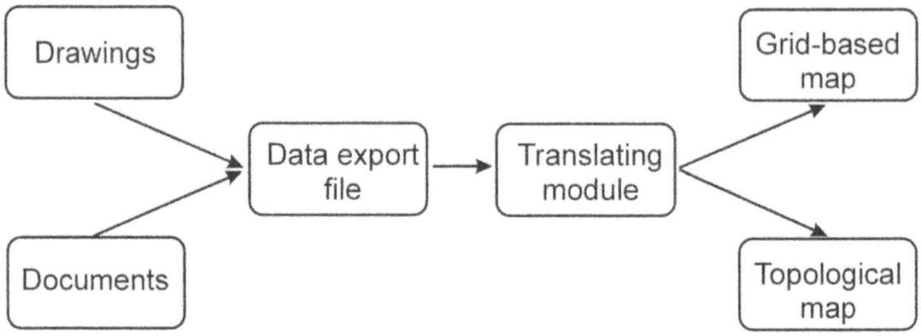

relation indicates the adjacency of objects, for example, spaces. When extending BIM toward MR applications, an additional relation describing accessibility of spaces should be added.

Let us consider now how one can generate MR-oriented maps taking the documentation in BIM format as the source. It is worth mentioning that the first attempts in this direction were undertaken one or two decades ago (Fennema et al. 1990; Murarka and Kuipers 2001). CAD drawings of that time contained merely the geometry of the building. Hence, there were no means to extract semantics out of them. In the papers by Borkowski et al., 2009 and Borkowski et al., 2010, we show how the transition of knowledge from BIM to MR can be accomplished. A schematic block-diagram of such process is depicted in Figure 4.

CAD files are usually written either in Drawing Interchange File (DXF) format or in Design Web Format (DWF). A dedicated parser transforms such files into MR formats, namely a grid-based occupancy map and a graph-based topological map. Figure 5 taken from Borkowski et al., 2009, shows the latter transformation. The hyper-graph depicted in Figure 5b describes the topology of floor-layout seen in Figure 5a.

BIM files are more valuable than CAD files, since they include information about the functional purpose of each part of the building. Such information is of considerable value for the mobile robot that performs certain tasks inside the building. It can be stored in attributes attached to nodes of the hyper-graph.

2.2. Outdoor Mapping

Mobile robots working outdoors are usually all-terrain vehicles covering large distances over land. Hence, their chassis has to be much more

Figure 5
Example of data transfer: a) part of floor layout; b) topological map (with permission of ASCE)

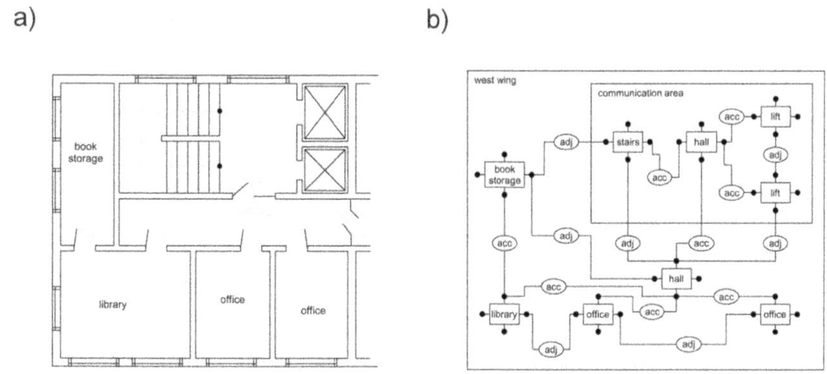

robust than the one for indoor service. On the other hand, the availability of maps is similar for both classes of robots. At present, the digital map of any place on Earth is either already available, for example, via Google Maps (Google Maps 2013), or can be obtained at reasonable cost in the form of aerial or satellite photographs. Miniature Unmanned Aerial Vehicles (UAVs) that have appeared in the market in the last decade cost even less. It is reported in the literature (Aeryon 2012) that fifty acres of land can be mapped within the accuracy of one inch in twelve minutes by means of the camera mounted on a tiny quadrocopter. Moreover, there exists an industry standard for retrieving, storing, manipulating, analyzing, and managing all types of geographical data. Pioneered in the 1960s in Canada, this standard is known as the Geographic Information System (GIS) (Longley et al. 2005).

The main feature of GIS maps is that they are composed of layers detailing such features as the shape of the terrain, roads and other communication networks, water resources, and so forth. GIS data are stored and made available for the user by a specialized Database Management System (DBMS).

The features relevant for a mobile robot depend on the type of application and on the character of the environment. Rescue robots or robotized agricultural machines work mainly in the open terrain. The main issue for them is the safe passage of the MR through a certain area. It depends on several factors, but the most significant is a slope of terrain. An MR can successfully maneuver through an area that has a slope with a measurement that falls within prescribed limits. For example, the RoMEGAT negotiates slopes inclined up to 15 degrees.

One of the layers of the GIS map indicates an elevation of terrain with respect to a certain reference level. Based upon this information, a topographical map can be rendered, where contour lines indicate the elevation. One of the major advantages of GIS is the possibility of performing a spatial analysis. Thus, a distribution of slope can be retrieved from the elevation layer and the areas which are impassable because of a steep slope can be marked on the map. Other reasons for restricted negotiability, like loose sand, swampy terrain or thick forest, can be also easily retrieved from the GIS data.

Robots working in urban areas can take advantage of the existing network of roads. Such a network can also be extracted from a proper GIS layer. However, the navigation on roads has been already successfully solved by the automotive industry. Commercial systems, like TomTom (2012) or Garmin (2012), use proprietary maps. However, street maps can be retrieved from free depositories in the Internet, for example, Open Street Map (OSM), (2012), and adopted for the purposes of the MR.

2.3. Maintenance of a Map

Let us consider now, how the information about the environment should be processed by an application-oriented mobile robot. Since an existing digital map can be taken as a ground truth, there is no need to build it by means of a tedious mapping process. Instead, such a map is taken from a depository, transferred to a format suitable for the mobile robot and stored in the memory of its controller.

In the case of an indoor robot, a map can be acquired once and the map would cover the entire working space (e.g., a building). Outdoor robots might be launched at different sites, sometimes separated by large distances from each other. In such circumstances, it might be better to keep the repository of maps on a server and to download a proper map to the robot prior to the start of its mission. This procedure can be efficiently implemented within a framework of Service-Oriented Architecture (SOA), as shown by Terlikowski et al. (2010).

Following a given map blindly is not wise, a fact which can be confirmed by anyone who uses the navigation system in a car. Fortunately, a mobile robot is able to inspect the environment with its sensors and to build a local map on the fly by means of the Simultaneous Localization and Mapping (SLAM) or other method. In principle, this should allow the robot to keep the map updated in accordance with measurements. However, practical implementation of the update procedure might pose certain problems.

Let us consider the situation when part of the stored map differs from that reconstructed on the basis of current observations. Figure 6 shows different types of possible discrepancies. It might happen that the two maps differ only in geometry, that is, they contain the same objects having more or less different dimensions and/or positions. Bearing in mind that the local map has been generated using the current position of the robot and that the estimate of this position is prone to errors, one should rather ignore slight geometrical discrepancies (Figure 6a) in favor of the ground-truth map.

Figure 6
Discrepancies between map and environment: a) negligible geometrical discrepancy; b) major geometrical discrepancy; c) topological discrepancy

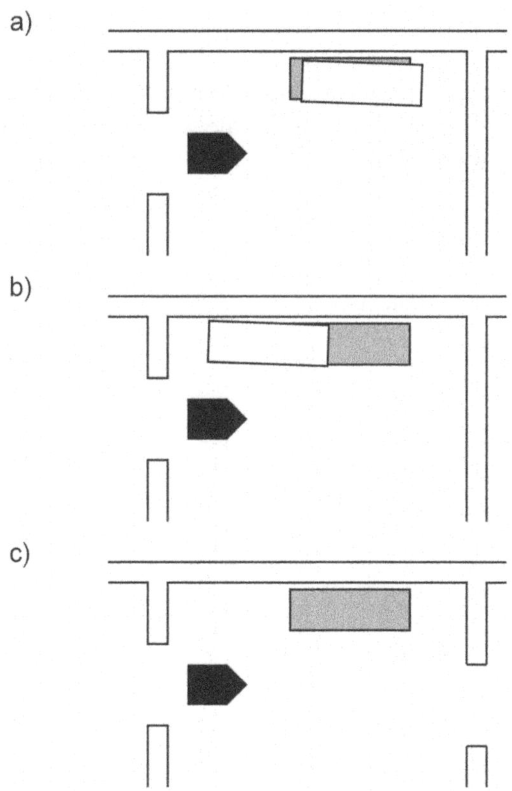

The situation shown in Figure 6b is different. The position of one object considerably differs from that stored in the map, whereas the rest of the observed environment stays in good agreement with the map. In this case, the map should be updated, possibly after approval by a human operator. A similar solution could be applied in the situation depicted in Figure 6c. Here, a new object—a door absent in the stored map—has been detected. This obviously should trigger an update of the map. Note that automotive navigation systems allow users to modify maps.

3. Self-Localization

Without considering flying robots that operate in 3D space, let us look at wheel-propelled units that will most probably prevail in the near future. Disregarding temporarily the slope of the terrain, we can project the working area of the robot onto a plane. Then, the current position of the robot is described by the components of vector $P = (X, Y, \Phi)$ as shown in Figure 7a. Here X and Y are the co-ordinates of the centre of robot in the global reference frame $\{0, X, Y\}$, whereas Φ is the direction toward which the robot is heading. Sometimes the shape of the robot and the restrictions imposed by non-holonomic[5] nature of its kinematics can be neglected. Then the robot is reduced to a point and $P = (X, Y)$.

Frequently it is necessary to estimate where a certain object is located relative to the robot. This can be done by introducing a local reference

Figure 7
Reference frames: a) global; b) local

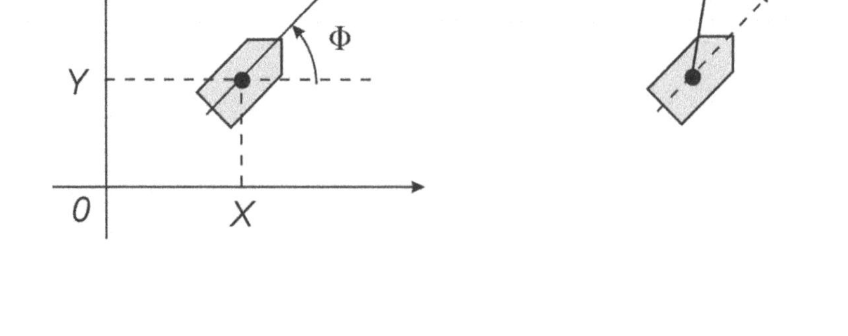

frame $\{0, \rho, \phi\}$, (Figure 7b). The relative position of object is described then by vector $p(\rho, \phi)$.

The position is a function of time, since the robot moves. Given the initial position, the robot can estimate its current position by: (a) dead reckoning;[6] (b) comparing observed environment with a given map; and (c) obtaining current coordinates from the GPS receiver. At present, most robots fuse together the probabilistic position estimates obtained from different sources. An exhaustive description of such a procedure can be found in Thrun et al. (2005).

3.1. Metric versus Topologic Localization

Describing the current position by means of numbers is called metric localization. This approach is quite natural in the case of stationary robots, where the head of the manipulator must be precisely positioned in space. The accuracy of localization for a mobile robot is of less importance. Hence, a qualitative or topological localization comes into play. In this approach, the digital coordinates are replaced by relations describing relative positions of objects in space. As an example, consider a part of the building shown in Figure 8a. This part consists of a corridor and two rooms A and B. It is convenient to describe it by a graph depicted in Figure 8b. The nodes of this graph correspond to particular spaces in the building, whereas the edges express relations between those spaces. Given such representation of the scene, we can describe the topological localization

Figure 8
Example of a part of building: a) floor plan; b) graph

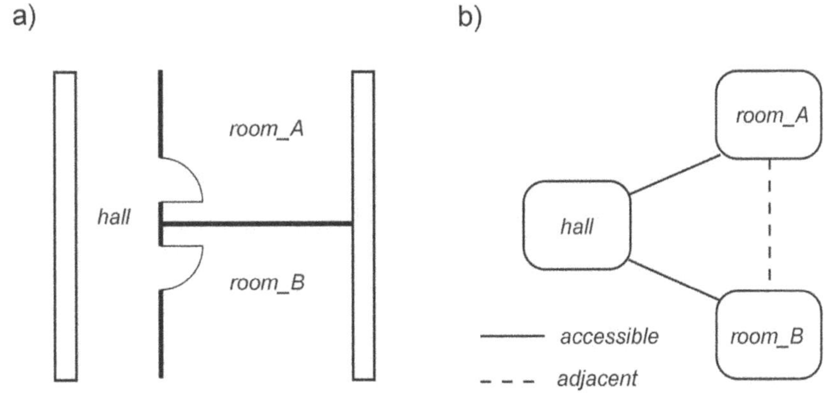

Figure 9
Example of an office building: a) overall view (courtesy of IPPT PAN); b) blocks

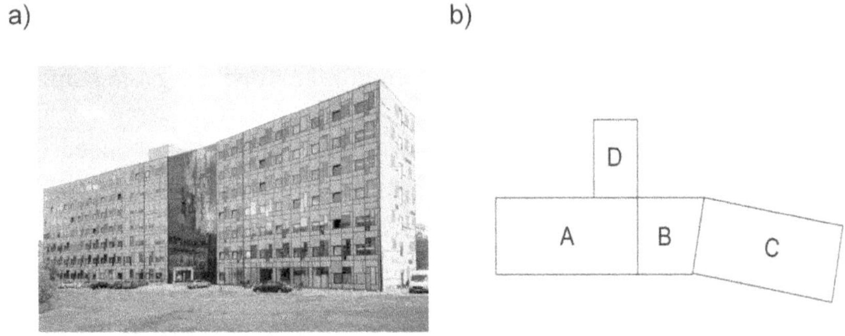

of an object by predicate[7] *localization (Place)*, where the argument *Place* can take values *corridor, room_A* or *room_B*. Both topological relations *adjacent (A, B)* and *accessible (A, B)* take places as their arguments but they differ in meaning. The adjacency means that two parts of the building have a common border, whereas the accessibility implies a direct connection, for example, a door, between them.

Note that topological localization allows the robot to navigate conveniently in the environments that possess a clear internal framework. Such is the infrastructure created by human beings: buildings, cities, roads, and so forth. Let us take an example of a multistoried office building shown in Figure 9a. A rough topological localization would simply indicate a block of the building (Figure 9b), like *localization (block_A)* or *localization (block_B)*. However, increasing the number of arguments of the predicate localization, we can easily expand the localization over the entire internal structure of the building. For example, the predicate *localization (block_C, floor_4, room_34)* refers to room 34 on the fourth floor of block C. In terms of the knowledge representation this requires the replacement of an ordinary graph by a hyper-graph,[8] as shown in Grabska et al. (2007).

Topological localization can be easily brought down to the level of a single part of the building. Thus, being inside a room, the robot may consider the zones *in_the_middle, at_the_wall, in_the_corner, at_the_door* or *at_the_window* (Figure 10a) as possible values of the argument *Place* of the predicate *localization*. In a corridor or a hall it is reasonable to distinguish the zone *at_the_door* (Figure 10b), since doors are the best landmarks in such spaces.

Figure 10
Topological localization inside building: a) in a room; b) in a corridor

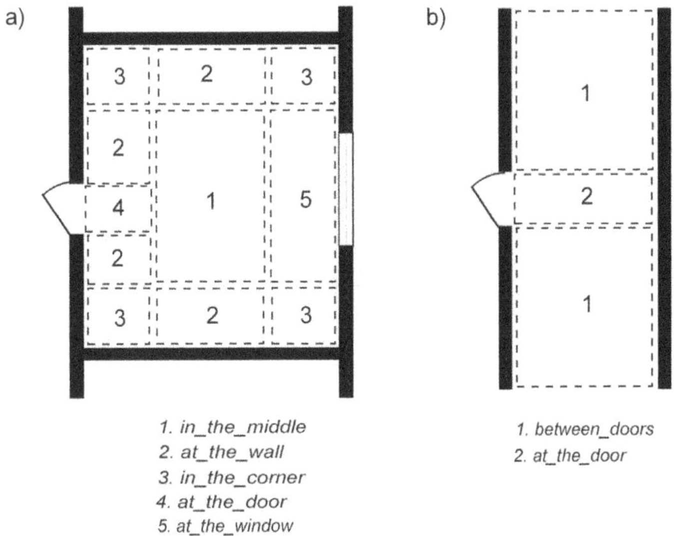

A position of an object, for example, an obstacle, relative to the robot can also be described topologically. Figure 11 shows the local reference frame suitable for this purpose. Obviously, the borders of zones *near, far, ahead, behind*, and so forth should be blurred rather than sharp. Reasoning over such entities would lead us away from classical first-order logic. The proper formalisms could be either fuzzy logic (Zadeh 1996) or a theory of rough sets (Pawlak 1991).

Topological localization considerably simplifies the interaction between human operator and mobile robot. We don't localize ourselves in terms of numbers; neither do other living species in nature. This should motivate us to use topological localization in MR more widely. However, completely abandoning the metric way of localization would be a mistake.

A reasonable solution might be mixed localization. Being in a room, the robot might know the symbolic descriptor of this room and, additionally, its own position $p(x, y, \varphi)$ related to a local frame of reference (Figure 12a). Such information could be stored in a predicate having four arguments: *localization(Place, x, y, φ)*. Localizing itself in a long corridor without doors and other landmarks is cumbersome, without taking advantage of the odometry or Inertial Measurement Unit (IMU). Hence, a proper predicate would be *localization(Place, x)*, as shown in Figure 12b.

Figure 11
Topological position of an object relative to robot

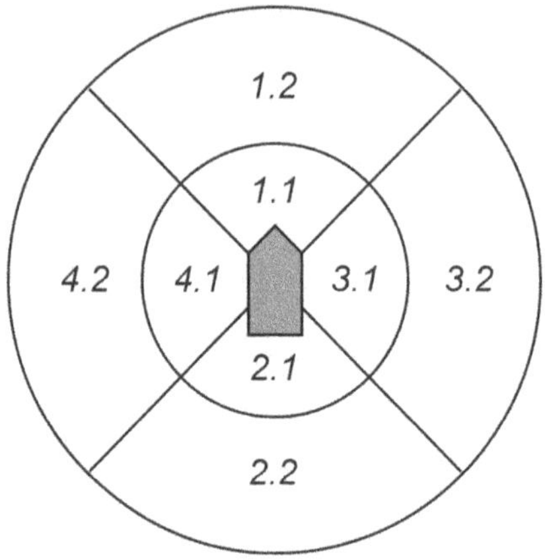

1.1. ahead_near	*3.1. to_the_right_near*
1.2. ahead_far	*3.2. to_the_right_far*
2.1. behind_near	*4.1. to_the_left_near*
2.2. behind_far	*4.2. to_the_left_far*

Figure 12
Mixed localization: a) in a room; b) in a corridor

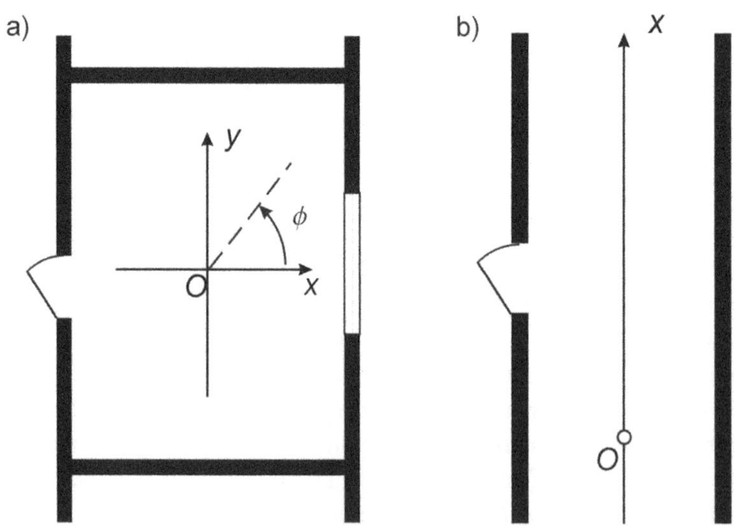

3.2. Indoor Localization

Given the progress in hardware and software for processing visual information, the present bias toward laser-based localization should shift in the future toward vision-based systems. Successful examples of such an approach can be already found in the literature, for example, Booij et al. (2009). Instead of relying only on odometry, inertial position tracking and on registration of patches of map obtained from laser range scanners, as was done until now, in future, one should increasingly take advantage of landmarks and sources of information placed inside a building. There is a broad palette of measures that make a building orientation-friendly in human terms. Among them are the floor plans and direction indicators placed at the crossroads, the walls painted in different colors, and the info labels attached to the doors. It is reasonable to assume that buildings meant to be served by mobile robots will be made localization-friendly for them. Let us discuss briefly how such a goal can be achieved.

The simplest way would be to retain a human-centered system of orientation landmarks and to force the mobile robot to use them. This would require the robot to detect printed labels inside the building and to read them. While in principle this is possible, this method is unlikely to be extensively used, since Optical Character Recognition (OCR) technology (Schantz 1982) is rather expensive. Employing cameras and lasers as reading devices, one would tend to simplify and standardize the pattern of information encoding. This has been done for years in industry with the help of Universal Product Code (UPC) barcodes (UPC 2013) and Quick Response (QR) codes (QR 2013).

The advantage of these codes lies in their speed and certitude. Usually, the twelve digits stored in a UPC code are not enough for MR purposes. The QR code stores information in a square matrix of black and white dots. The storage capacity depends on the resolution of this matrix. A code with a resolution of 177 x 177 would store over 4000 characters but it would be cumbersome to read from far away. Hence, a resolution of 21 x 21 or 33 x 33 is preferable. Figure 13 shows the label *Room 4.34* written by the author using this format.

A common disadvantage of vision-based labels is that they might be occluded or not clearly visible. The robot has to be very close when it passes the label. Therefore, Radio-Frequency Identification (RFID) tags can be considered as an alternative solution (Glover and Bhatt 2006). These are tiny chips containing radio transmitter-receiver and memory. After receiving a request from the transmitter mounted on mobile robot,

Figure 13
QR–code storing room label

the RFID tag would send back the information stored in its memory. Passive tags do not have their own batteries—they use the energy contained in the request. They are extremely cheap (about 15 US cents/tag) but can operate only up to a short distance of two meters. Active and semi-active tags are equipped with batteries and cover a distance of up to 200 meters. It is estimated that new tags working in the range of 3–10 GHz will cost about US$5 apiece.

The first attempts to apply RFID tags in MR have already been reported in the literature (Schneegans et al. 2007; Rohweder et al. 2009). The main difficulty in using them as localization aids is that RFID technology does not provide the azimuth of reading. Hence, when the robot receives responses from multiple tags, it is difficult for it to localize the relevant objects. This can be overcome by including position in the information stored on the tag.

3.3. Outdoor Localization

Working in open terrain, a mobile robot can take advantage of a GPS fix (see details in, e.g., Kaplan and Hegarty, 2006). Provided at least four satellites are working simultaneously, the GPS allows the robot to obtain its current position in 3D reference frame: latitude, longitude, and altitude. The accuracy of localization depends upon many factors. On an average, the latitude and the longitude can be estimated to be within ±5 meters. The error in altitude is usually bigger, but this is not significant for most applications.

Experience gained so far in automotive navigation systems suggests that GPS-based localization is accurate enough for driving on roads. Mobile robots that operate in open terrain can take the GPS fix as an initial estimate of position. This estimate can be enhanced, if necessary, by measuring distances to landmarks of known positions.

4. Path-Planning

In general, planning a collision-free path for mobile robots can be seen as a problem of exercising optimal control: find a trajectory in the configuration space that meets given constraints, that is optimal in a prescribed sense and that brings the robot from a known initial state to a desired final state. The configuration space is spanned over control variables like the torques applied to the driving wheels. The constraints may be two-fold in nature: those imposed by the kinematics of the robot and other imposed by its environment. An example of kinematic constraint is that a wheeled robot cannot move in the direction parallel to the axis of the wheels. The simplest and yet the most significant environmental constraint is that the robot not collide with obstacles. The optimality criterion is usually either a shortest path criterion or a minimum energy consumption criterion.

Path-planning is comprehensively described in the monographs written by Jean-Claude Latombe (1991) and Steven LaValle (2006). Hence, in Chapter 2, we restrict ourselves to the simplified, yet oft- used formulation of the problem. In this formulation, the mobile robot is treated as a point on the 2D plane. Kinematic constraints are disregarded and the dimensions of the robot are taken into account by expanding obstacles in such a way that a collision-free passage is warranted. Under such assumptions, path planning means: find a trajectory *T* that leads the robot in a collision free way from the current position *S* to the goal position *G*, as shown in Figure 14. It is assumed that *S* is supplied by the self-localization procedure, whereas *G* is given by the task planner.

Figure 14
Basic notions in path–planning

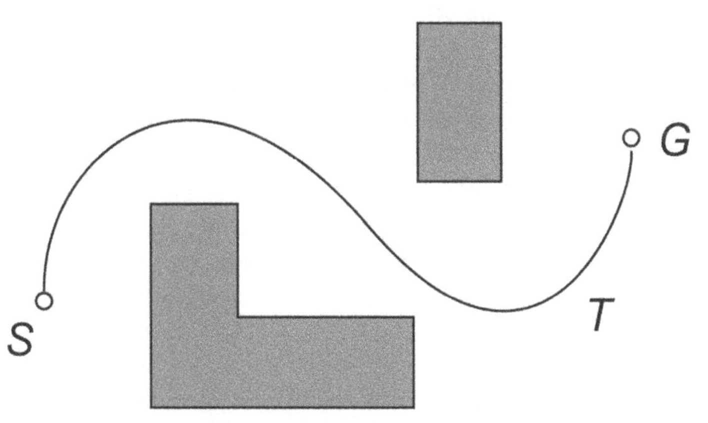

Figure 15
Visibility graph–based path–planners: a) based upon vertices of obstacles; b) based upon equidistance points

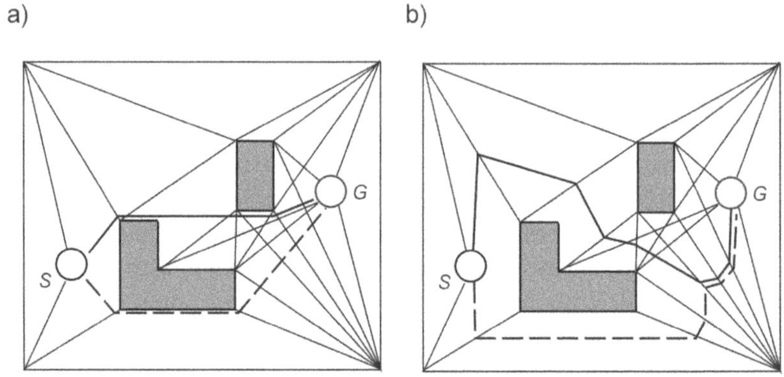

Algorithms solving the above-stated problem can be divided into three classes. The first class is based on pure geometry. The working area W of the robot is considered to be a sum of free space F and space occupied by obstacles O: $W = F \cup O$. Let obstacles be polygonal and let V be a set of vertices describing them. Then a visibility graph can be spanned over the set $\{P, G, V\}$ (Figure 15a). The edges of this graph connect points that are visible to all of them. With the help of this graph, we can find a shortest path leading from S to G. Many efficient algorithms are available in the theory of graphs for solving such problems. The most well-known of them is the Dijkstra algorithm (Dijkstra 1959).

A disadvantage of the procedure shown in Figure 15a is that it leads the robot close to the obstacles. This can be mitigated by spanning the visibility graph over equidistant points between vertices, as shown in Figure 15b.

Graph-based planning is quick, but it generates piecewise linear trajectories with sharp turns at vertices. Such trajectories must be smoothened for the robot to execute them. An alternative is to use potential field-based planning, which generates smooth trajectories. This approach is based on Physics. The robot is treated as a particle moving in an electrostatic field generated by the goal and the obstacles. The robot and the goal are given opposite charges, so they attract each other. The obstacles are loaded with the same charge as the robot. This introduces repellent forces between them. Solving the equation of the motion of a particle, one obtains a smooth trajectory avoiding obstacles and leading the robot

Figure 16
Potential field–based path–planning: a) basic concept; b) example of failure

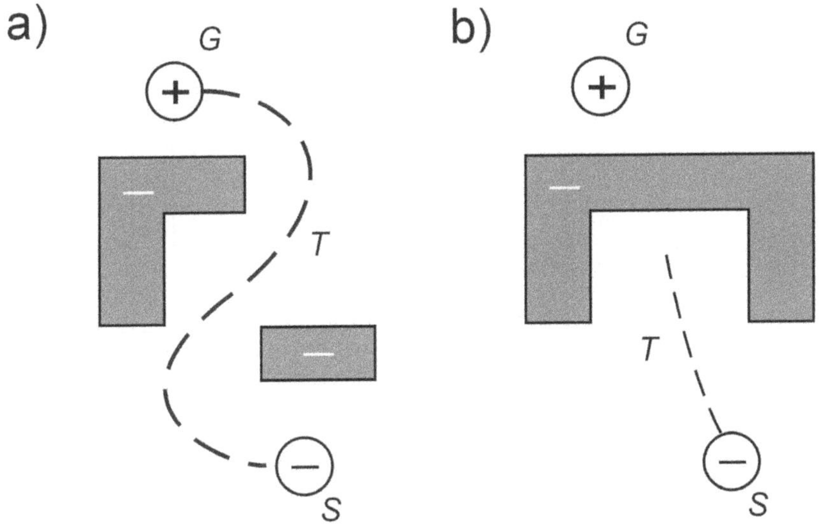

to the goal (Figure 16a). A disadvantage of this method is that the robot can get trapped, as shown in Figure 16b.

The third idea widely used in path planning is based upon the propagation of a wave over the environment. This method has been "re-invented" by several authors, who used different analogies from Physics and gave essentially the same method, different names. Let us stick to "diffusion-based path planning" proposed by Luc Steels (1990). The approach is illustrated in Figure 17.

The diffusion-based path planner works on a grid map. It starts adding "height" to free cells that surround the goal. The process is stopped when the front of the diffusion reaches the cell occupied by the robot. The result is a "hill" with the goal cell on the top of it. This can be envisioned as the result of pouring water from an outlet located at the goal cell and letting it freeze under low temperature. The second phase of the path-planning procedure is simple: one follows the steepest gradient of the hill to reach the goal. Diffusion-based path planning is simple to implement, it avoids blind alleys and can be easily extended to dynamic situations, when the goal position ceases to be stationary. Its disadvantage lies in the need for a sufficiently high-resolution map, which might be cumbersome for larger areas.

Figure 17
Diffusion–based path-planning

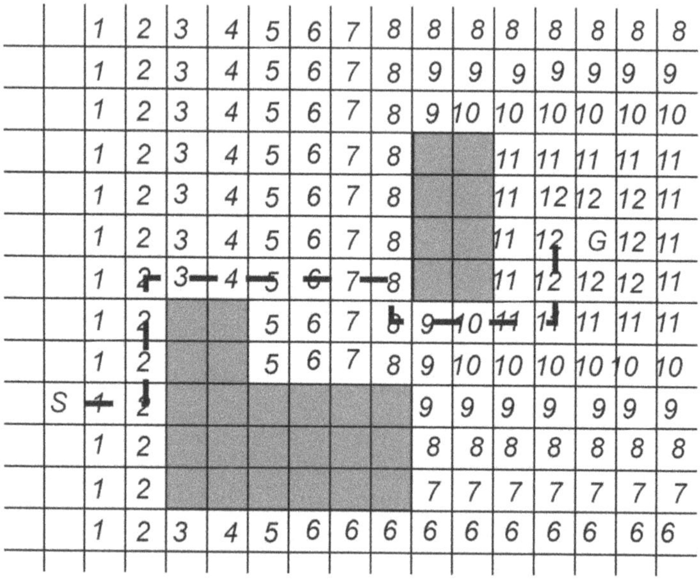

4.1. Indoor Path Planning

The floors of most public buildings have a simple structure that is identical on each storey. A mobile robot designed to serve in such a building should be equipped with a palette of motion primitives, like driving along a corridor, entering a door, or using an elevator. Combined with a topological or a hybrid map, such primitives allow fast and efficient navigation inside a building. Driving along a corridor, for example, can be achieved by a controller that minimizes the difference between the distances measured by the robot to the left and the right wall. Opening, entering, and then closing a door—trivial tasks for a human—pose serious challenges for robots. It is probable that robot-friendly buildings will have remotely operated sliding doors. A similar solution can be adopted for elevators, in order to eliminate push-buttons. It works already in commercially available tugs that transport goods in hospitals (Aethon 2012).

It is clear that a robot working inside a building runs the same path many times. Hence, instead of planning the path from scratch each time when a certain task is to be performed, the robot should try to utilize one of the pre-planned routes. For example, a social robot deployed in a

Figure 18
Alternative routes in a single family house

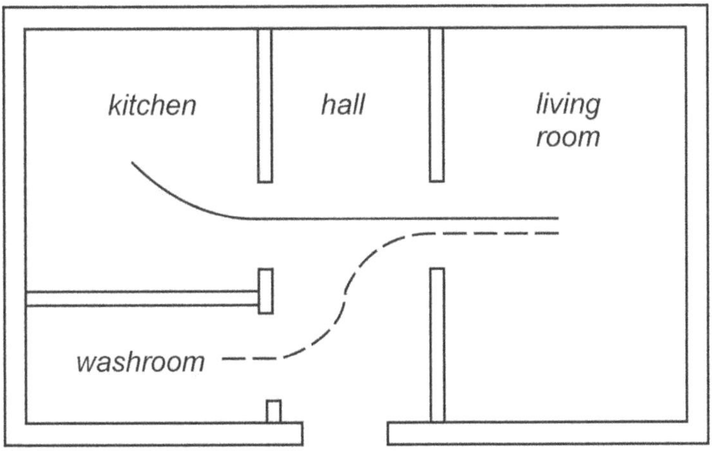

single-family home should know in advance how to reach a kitchen or a bathroom from a living room (Figure 18).

Such pre-planned trajectories could lead the robot to a central point in the relevant room. Further portions of the path are task-dependant; it could be planned by means of, say, the diffusion method (Figure 17).

Taking advantage of a network of pre-planned trajectories is a standard solution in industrial applications, for instance, in computerized warehouses. The means of controlling multiple robots over such a network is shown in Roszkowska et al. (2007).

4.2. Outdoor Path Planning

Mobile robots that drive on roads face the same problem in path planning as humans driving cars. Hence, there is no need to invent path-planning procedures for such robots. Instead, the path planners used in automotive navigation systems can be adjusted to serve in the MR domain. The necessary changes are related mainly to the interface. The navigation system of a robot becomes the goal assigned by the task-planner, whereas the navigation system of a car gets it from the driver.

The situation becomes different for tasks that require off-road driving. For a single mobile robot, any of the path-planning procedures mentioned in the introductory part of this section can be applied. The diffusion method offers two advantages: first, it can take into account differences

Figure 19
Typical formations of multiple mobile robots: a) a platoon; b) a sweeping line; c) a triangle

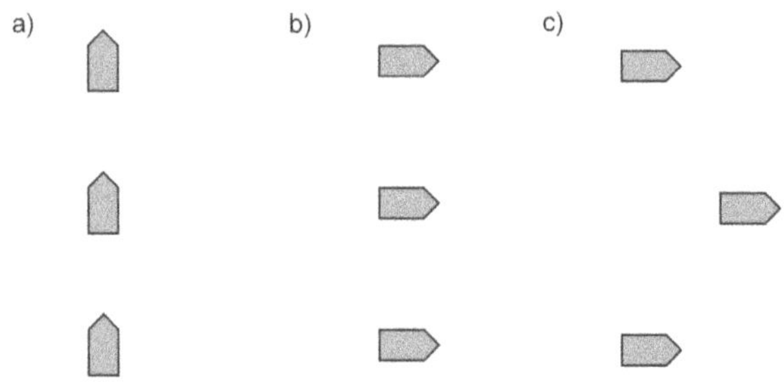

in the negotiability of the terrain; second, it can be expanded in case multiple mobile robots are needed. Examples of applications that require co-operation of multiple mobile robots can be found in many areas. In agriculture these would be the autonomous harvesters, in forestry, the wood-cutters, in the military, the mine-sweepers.

A common feature of multi-robot navigation is that robots working together should keep to certain patterns of motion. Typical formations of this kind are depicted in Figure 19. The orchestrated motion of multiple robots has been studied for a few decades and several path-planning techniques were proposed. One of them is based upon a concept of Cellular Neural Network (CNN), for example, Przybylski and Siemiątkowska (2012).

CNN is a dynamic system composed of cells connected in a network (see details in Chua and Roska, 2002). The word "neural" in this computational paradigm indicates the relation with the commonly known concept of the Artificial Neural Network (ANN). The difference between ANN and CNN lies in the range of interaction between neurons and cells. In the ANN model, an excitation flows from the input layer to the output layer, whereas in the CNN model, it is limited to a properly defined neighborhood of a cell. In other words, CNN works like a recurrent ANN. It receives an excitation, undergoes a transition state, and stabilizes at a certain attraction point that corresponds to the solution.

CNNs are well-suited for storing grid-based maps, because each cell of the map can be assigned directly to a cell in the cellular network. Moreover, it is possible to implement a multi-layer CNN model that stores not only the occupancy of individual cells but other features of the terrain as well. Thus, a CNN based path planner can take into account differences in the negotiability of the terrain and generate a route that optimizes the energy consumed and keeps it to a minimum.

5. Final Remarks

Intelligent mobile robots seem today to have reached the level of maturity that allows them to leave research labs and enter applications. In order to accomplish this transition, several paradigms used so far by the MR community should be evaluated critically and possibly modified. One of those paradigms is the concept of autonomous navigation. The following statements summarize our conclusions:

- Instead of building maps from scratch, mobile robots should take them from depositories and update according to the observed reality.
- Topological or hybrid localization should be preferred whenever possible. Outdoors, the advantage of direct GPS fix should be fully exploited.
- Pre-planned routes should be used, instead of planning a route each time from scratch.

We believe that the remarks and proposals presented in this chapter will help designers to develop good—meaning efficient, robust and affordable—mobile robots in the future. Another critical issue is to make such robots user-friendly and psychologically acceptable to ordinary people who are far from being familiar with computerized machines. This aspect has been discussed in other chapters of this book.

Notes

1. Institute of Technical Cybernetics renamed in 2005 as the Institute of Computer Science, Automation and Robotics, Wrocław University of Technology, Wybrzeże Wyspiańskiego 27, 50–370 Wrocław, Poland.
2. Laboratory of Adaptive Systems renamed in 2000 as the Laboratory of Intelligent Systems as of 2013, Institute of Fundamental Technological Research, Polish Academy of Sciences, ul. Pawińskiego 5B; 02–106 Warsaw, Poland.
3. Materials Engineers Group Ltd, ul. Wołoska 141, 02–507 Warsaw, Poland.
4. This acronym should stand for Sequential Localization and Mapping rather than Simultaneous Localization and Mapping, since it is not possible for a robot to draw a map and to localize itself at the same time. Such an attempt would resemble the achievements of Baron Münchausen who escaped from a swamp by pulling his own hair! In fact, when using the SLAM procedure, one first estimates the current position and then builds new segments of the map.
5. Non-holonomic constraint for a wheel-propelled robot simply means that such a robot cannot move in the direction of the wheel axis.
6. Dead reckoning is the process of calculating current position by using a previously determined position and modifying it according to a known or estimated displace-

ment. Its simplest version is odometry, where the displacement of the robot is calculated by counting the rotations of its wheels.
7. Predicate in first-order logic is a function that converts its arguments into logical true and false values.
8. Hyper-graph is a graph having sub-graphs as its nodes.

References

Aeryon. http://www.aeryon.com/news/pressreleases/248-softwareversion5.html, retrieved November 17, 2012.

Aethon. www.aethon.com/solutions/deliver, retrieved November 5, 2012.

Ambroszkiewicz, S., A. Borkowski, K. Cetnarowicz, and C. Zieliński, eds. *Intelligence around Us: Co-operation of Software Agents, Robots and Intelligent Devices* (in Polish). Warsaw: Academic Publishing House EXIT, 2010.

Booij, O., Z. Zivkovic, and B. Kröse. "Efficient Data Association for View Based SLAM Using Connected Dominating Sets." *Robotics and Autonomous Systems* 57, no. 12 (2009): 1225–34.

Borkowski, A., E. Grabska, and W. Palacz. "Modeling Buildings for Mobile Robots." In *Proceedings of ASCE/EG–ICE International Workshop on Computing in Civil Engineering*, ed. C. H. Caldas and W. J. O'Brien, 11–21. Austin, TX: ASCE Publication, 2009.

Borkowski, A., G. Engles, C. Lewerentz, W. Schäfer, A. Schürr, and B. Westfechtel, eds. "Towards Semantic Navigation in Mobile Robotics." *Graph Transformations and Model–Driven Engineering*, LNCS 5765, 719–48. Berlin, Heidelberg: Springer, 2010.

Chua, L., and T. Roska. *Cellular Neural Networks and Visual Computing: Foundations and Applications*. Cambridge: Cambridge University Press, 2002.

Curiosity. www.nasa.gov/mission_pages/msl, retrieved December 18, 2012.

Dijkstra, E. W. "A Note on Two Problems in Connection with Graphs." *Numerische Mathematik* 1 (1959): 269–71.

Eastman, C., P. Teicholz, R. Sacks, and K. Liston. *BIM Handbook: A Guide to Building Information Modeling for Owners*. Hoboken: Wiley, 2008.

Faderewski, M., S. Ambroszkiewicz, A. Borkowski, K. Cetnarowicz, and C. Zieliński, eds. "Representing Environment of Mobile Robot" (in Polish), "Intelligence Around Us: Co-operation of Software Agents," *Robots and Intelligent Devices* (in Polish), 13–42. EXIT Academic Publishing House, Warsaw, 2010.

Fennema, C., et al. "Model-Directed Mobile Robot Navigation." *IEEE Transactions on Systems, Man and Cybernetics* 20 (1990): 1352–69.

Garmin. www.garmin.com, retrieved October 8, 2012.

Glover, B., and H. Bhatt. *RFID Essentials*. Sebastopol, CA: O'Reilly Media, Inc., 2006.

Google Maps. www.maps.google.com, retrieved January 10, 2013.

Grabska, E., G. Ślusarczyk, M. Glogaza, and M. Kurzyński, eds. "Design Description Hyper-Graph Language." In *Computer Recognition Systems 2, Advances in Soft Computing*, vol. 45, Berlin, 2007.

Kaplan, E. D., and C. J. Hegarty. *Understanding GPS—Principles and Applications*. Norwood, MA: Artech House, 2006.

Kinect. www.theverge.com/products/kinect-for-windows/4833, retrieved November 11, 2012.

Latombe, J. C. *Robot Motion Planning*. Boston, MA: Kluwer, 1991.

LaValle, S. M. *Planning Algorithms*. Cambridge MA: Cambridge University Press, 2006.

Longley, P. A., M. F. Goodchild, D. J. Maguire, and D. W. Rhind. *Geographic Information Systems and Science*. Chichester: Wiley, 2005.

Mower. www.robotlawnmowers.co.uk, retrieved November 11, 2010.

MTi. www.xsens.com, retrieved November 11, 2012.

Murarka, A., and B. Kuipers. "Using CAD Drawings for Robot Navigation." *IEEE Transactions on Systems, Man and Cybernetics* 2 (2001): 678–83.

OSM. www.openstreetmap.org, retrieved November 17, 2012.

Parker, L. E. web.eecs.utk.edu/~parker/Courses/CS594-fall08/Lectures/Aug-28-History.pdf, retrieved 12 November, 2012.

Pawlak, Z. *Rough Sets: Theoretical Aspects of Reasoning about Data*. Kluwer: Academic Publishing, Dordrecht, 1991.

Predator. www.af.mil/information/factsheets/factsheet.asp?fsID=122, retrieved December 18, 2012.

Przybylski, M., and B. Siemiątkowska. "A New CNN-Based Method of Path Planning in Dynamic Environment." *Lecture Notes in Artificial Intelligence* 2 (2012): 484–92.

QR–code. http://www.qrcode.com/en/qrfeature.html, retrieved 16 January, 2013.

RoboToys. www.robotoys.com, retrieved November 11, 2012.

Rohweder, K., P. Vorst, A. Zell, I. Petrovic, and A. J. Lilienthal, eds. "Improved Mapping of RFID Tags by Fusion with Spatial Structure." In *Proceedings of 4th European Conference on Mobile Robots (ECMR'09)*, 247–52. Mlini/Dubrovnik, Croatia: KoREMA, 2009.

Roomba. http://www.irobot.com/us/robots/home/roomba.aspx, retrieved January 13, 2013.

ROS. www.ros.org, retrieved November 11, 2012.

Roszkowska, E., B. Kreczmer, A. Borkowski, M. Gnatowski, W. Burgard, and H.-M. Gross, eds. "Distributed supervisory Control for a System of Path-Network Sharing Mobile Robots." In *Proceedings of 3rd European Conference on Mobile Robots (ECMR'07)*, http://ecmr07.informatik.uni-freiburg.de/proceedings/ECMR07_0056.pdf, retrieved November 17, 2007.

Schantz, H. *The History of OCR*. Manchester Center, VT: Recognition Technologies Users Association, 1982.

Schenck, D. A., and P. R. Wilson. *Information Modeling the EXPRESS Way*. New York: Oxford University Press, 1994.

Schneegans, S., P. Vorst, A. Zell, W. Burgard, H.-M. Gross, eds. "Using RFID Snapshots for Mobile Robot Self-Localization." In *Proceedings of 3rd European Conference on Mobile Robots (ECMR'07)*, http://ecmr07.informatik.uni-freiburg.de/proceedings/ECMR07_0026.pdf, retrieved November 22, 2007.

Sick. www.mysick.com/products, retrieved November 11, 2012.

Siemiątkowska, B., A. Borkowski, R. Chojecki, M. Gnatowski, W. Mokrzycki, and J. Szklarski. *Representing Environment of Mobile Robot* [in Polish]. Warsaw: EXIT Academic Publishing House, 2011.

Steels, L. "Exploiting Analogical Representations." *Journal of Robotics and Autonomous Systems* 6, no. 1 (1990): 71–88.

Terlikowski, G., S. Ambroszkiewicz, A. Borkowski, K. Cetnarowicz, and C. Zieliński, eds. "SOA-Based Architecture of Mobile Robot." (in Polish) *Intelligence around Us: Co-operation of Software Agents, Robots and Intelligent Devices* (in Polish), 43–86. Warszaw: EXIT Academic Publishing House, 2010.

Thrun, S., W. Burgard, and D. Fox. *Probabilistic Robotics*. Boston: MIT-Press, 2005.

TomTom. www.tomtom.com, retrieved October 8, 2012.

UPC–code. http://www.gs1us.org/resources/standards/ean-upc, retrieved 16 January, 2013.

Vigilant. www.vigilantrobots.com, retrieved November 11, 2012.

Walter, W. G. "An Imitation of Life." *Scientific American* 182 (1950): 42–45.

Wołczowski, A., D. E. Grierson, G. Rzevski, and R. A. Adey, eds. "Autonomous Mobile Robot Ulisses." *Applications of artificial intelligence in engineering VII,* 883–94. Southampton: Computational Mechanics Publishers, 1992.

Xsens. www.xsens.com, retrieved November 17, 2012.

Zadeh, L. A., G. J. Klir, and Bo Yuan, eds. "Fuzzy Sets, Fuzzy Logic and Fuzzy Systems." *Advances in Fuzzy Systems—Applications and Theory*, vol. 6. Singapore: World Scientific Press, 1996.

Locomotion Mode Transition Control of a Multi-Locomotion Robot

Toshio Fukuda and Kosuke Sekiyama
Department of Micro-Nano Systems Engineering
Nagoya University

Tadayoshi Aoyama
Department of System Cybernetics
Hiroshima University

Yasuhisa Hasegawa
Department of Intelligent Interaction Technologies
University of Tsukuba

Qiang Huang
School of Mechatronical Engineering
Beijing Institute of Technology

1. Introduction

With the rapid progress of robotic technologies, robots are today expected to accomplish a number of tasks that human beings find tedious or dangerous. In this background, industrial robots were developed to perform tasks that were physically laborious for humans, such as the conveyance of heavy materials or to implement tough jobs such as mindless jobs. Robots are also desirable substitutes for human workers in hazardous places, such as nuclear power plants or outer space.

In Japan, a declining birth rate and the consequent jump in the population of the aged, has led to a dire social situation. Not only is there a shortage of productive manpower but also a serious lack of nursing care for the old. To solve these problems and to enrich people's lives, robots are necessary. There is an urgent requirement for life-supporting robots equipped with high mobility and intelligent systems that recognize the various situations and environments encountered in daily human life, and that are able to adapt and react to them quickly.

The idea of building machines as substitutes for living creatures has been around for a very long time. For example, in the 15th century, Leonardo da Vinci drew bird-like flying machines. Nowadays, the idea has become a reality and robots that utilize locomotion techniques similar to those used by human beings and other animals have been developed (Saito and Fukuda 1997). Locomotion robots that employ walking, flying, or swimming can be said to have been inspired by legged animals, birds, and fish. These robots should technically be called bio-inspired robots. However, Randall Beer et al. (1997) first used the term "biologically inspired" in the title of a journal article in 1997. In this essay we have reviewed their efforts to incorporate some of the principles of walking in animals into the design and control of legged robots. Life-supporting robots are an extension of bio-inspired robots considering that the area of action for life-supporting robots is our living environment. However, most bio-inspired robotic research deals with a single task; thus, the environments the robots can adapt to are limited because they are designed to specialize in doing a single task.

Here, focusing on the difference between conventional bio-inspired robots and living creatures on Earth, it is found that living creatures have great "diversity of locomotion." Living creatures have various forms of locomotion and are capable of switching between forms depending on their surroundings, situations, and purposes. For example, gorillas can achieve high mobility in forests by switching between bipedal walking, quadrupedal walking, climbing and brachiation, depending on the circumstances. "Diversity of locomotion" that living creatures have cannot be seen in previous bio-inspired robotic research. It is one of the most important issues in putting life-supporting robots to practical use.

Keeping the preceding discussion in mind, we have proposed the Multi-Locomotion Robot (MLR) as a novel bio-inspired robotic concept (Kajima et al. 2004; Aoyama, Hasegawa, et al. 2009; Aoyama, Sekiyama, et al. 2009; Lu et al. 2012). The MLR can perform several types of locomotion and can choose the proper one on an as-needed basis in order to function as the life-supporting robots that can assist humans in their daily lives. The following section introduces the concept of MLR and explains how the robot can choose between bipedal and quadrupedal walk according to suitability, walking stability, and efficiency.

2. Multi-Locomotion Robot

Diversity of locomotion, which has not been dealt with in previous work on bio-inspired robotics, is one of the significant points that researchers have to focus on so as to put life-supporting robots into practical use like a human partner. An intermediate physical structure between primates such as chimpanzee, orangutan, and human seems to be most appropriate in evolutionary term, when we consider what necessary mechanism and locomotion form is in order to accomplish human tasks (Eimerl and DeVore 1966; Simons 1972).

The new concept of a bio-inspired robot, the multi-locomotion robot, is shown in Figure 1. The multi-locomotion robot is a novel bio-inspired robot which can perform several kinds of locomotion such as bipedal walking, quadrupedal walking, and brachiation.

2.1. Bipedal Walking

In all living creatures with limbs in nature, the animals that exhibit bipedal walking are mammals, such as humans and anthropoids closely related to humans (human, chimpanzee, orangutan, gibbon, and gorilla). These creatures acquired bipedal locomotion because of the evolutionary process. Bipedal walking freed them to use their hands for other tasks. Consequently, it is thought that their intelligence is superior to other animals and they possess an excellent ability to adapt to a dynamic environment.

Figure 1
Concept of the Multi-Locomotion Robot

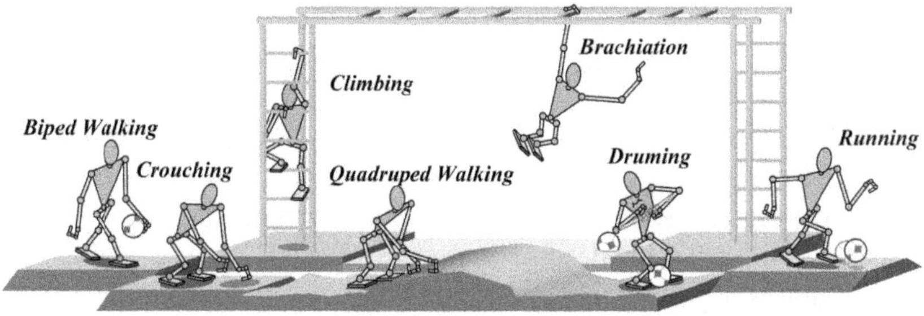

The model of bipedal walking is normally an inverted pendulum (Donelan et al. 2002). The dynamics of an inverted pendulum are intrinsically unstable, hence in order to achieve bipedal walking, it is necessary to control the body dynamics skillfully so as to compensate for instability. However, the Center of Mass (COM) of bipedal locomotion is kept on a higher position than the COM of quadrupedal locomotion or creeping; thus, it has merit that the energy efficiency of moving could be high by taking advantage of the gravity effects. Also, bipedal walking posture can get large measuring range of external environment due to the high position of bipedal posture.

Since most routine human tasks are conducted in a bipedal posture, the necessity for a robot to use bipedal walking is natural. In addition, the residential environment is designed to fit the human body structure. That is the reason why bipedal locomotion is suited for life-supporting robots working there. Recently, there has been lots of active research aiming to accomplish bipedal walking with humanoid robots. It is one of the most exciting areas of research in recent years. Besides, accomplishment of faster locomotion by developing running robots is on the anvil.

2.2. Quadruped Walking

Animals performing quadruped locomotion are familiar to us (Muybridge 1957). Quadruped walking is the locomotion mode that many of the terrestrial animals employ. It is known that there are several gait patterns in quadruped walking such as crawl gait (Figure 2(1)), pace gait (Figure 2(2)), trot gait (Figure 2(3)), bounce gait (Figure 2(4)), and gallop gait (Figure 2(5)) (Formal'sky et al. 2000). Crawl gait has periods of three-feet contact and four-feet contact; at least three feet are in contact with the ground at any time. Thus, the walking stability is quite high; however, it is not suitable for fast locomotion in terms of energy efficiency.

Most mammals, such as cats, dogs, and horses, have a particular gait at low velocity, in which the body is supported constantly by two legs on the ground. Each leg moves forward in the following order: right rear-leg, right front-leg, left rear-leg, left front-leg. As the walking velocity increases, the number of supporting legs in contact with the ground decreases, especially in trot gait (Herr and McMahon 2001a). and pace gait. In trot gait, the legs at the opposite corners—that is, the feet of the right rear-leg and left front-leg—contact the ground simultaneously by turns. On the other hand, in pace gait, the feet of the legs on the same side—that is, right rear-leg and right front-leg—are in contact with the

Figure 2
Quadruped Gait Graphs. Stance Phases are Drawn as Bars.

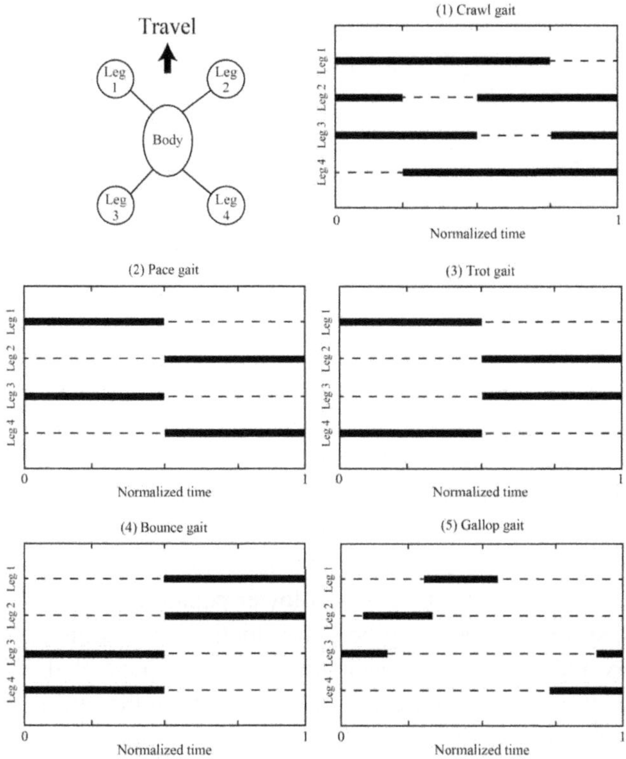

ground at the same time. At the highest velocity, namely, gallop gait, neither of the four feet are in contact with the ground for regular periods (Herr and McMahon 2001b). It is known that animals gain high energy efficient locomotion by switching these gait patterns according to walking velocity or state (Hoyt and Taylor 1981).

Quadruped locomotion is more stable than bipedal walking, hence it can be said that quadruped locomotion is suited for ambulation in an environment with many obstacles or rough and rugged terrain. In order to put partner robots into practical use, robots need to be able to ambulate on not only flat ground but also on irregular ground. Hence, in the environment where bipedal locomotion is unstable, robots should choose quadrupedal walking. Also, in the circumstances where the degree of spatial freedom is limited such as a narrow cave or tunnel, quadruped walking is effective.

Figure 3
Brachiation of a Gibbon

2.3. Brachiation

Brachiation is one of the modes of locomotion adopted by arboreal animals such as gibbons and orangutans. In this locomotion pattern, the animals dangle on the branch and swing from branch to branch like a pendulum as shown in Figure 3. It is known that the dynamics of brachiation are shown pendulum (Bertram et al. 1999; Bertram and Chang 2001). In locomotion on the ground, the COM is higher than the point that supports its weight, whereas in brachiation, the movement is performed dangling on the branch and the COM is in a lower position. Chang et al. (2000) built the system to observe gibbon brachiation, and estimated the COM trajectory and alteration of mechanical energy (Chang et al. 2000).

Brachiation enhances the ambulating ability of the multi-locomotion robot. It is difficult to perform bipedal or quadrupedal locomotion on ground which is not firm. However, if there is something to hold on and dangle from, such as a branch or a pipe, it is possible to ambulate using these.

2.4. Gorilla Robot III

We built and developed Gorilla Robot III as the prototype of MLR. The overview and link structure of Gorilla Robot III is shown in Figure 4. Its height is about 1.0 m and weight is about 24.0 kg. The mechanical structure is designed as follows: 6 degree of freedom (DOF) leg, 5 DOF arm, 2 DOF lumbar. Each joint is actuated by an AC servo motor. The computer, AD/DA board, counter board, and power, are fixed outside of the robot. A laser rangefinder is installed at the neck of the robot as a sensor for recognition of a slope (see Figure 5). Its angular resolution is 0.36 degress, scan angular range is 240 degrees, scan time is 100 m/s, and maximum range of detection is 4.0 meters. The rotation axes of the

Figure 4
Gorilla Robot III

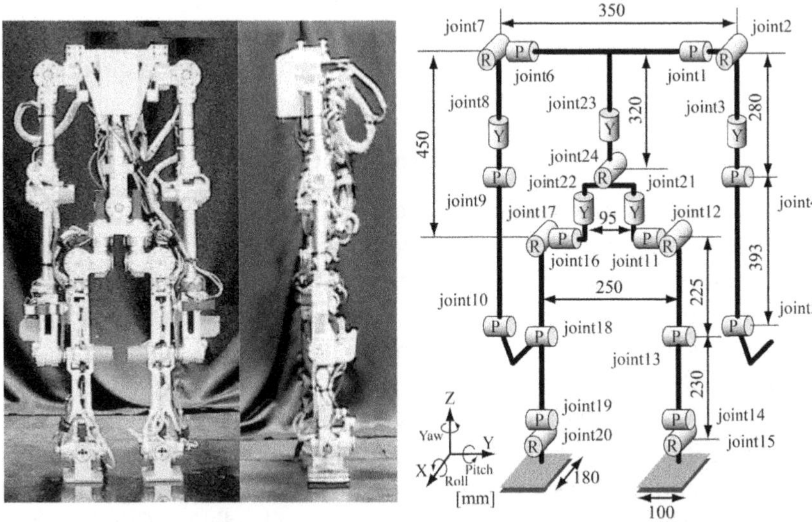

Figure 5
Laser Rangefinder

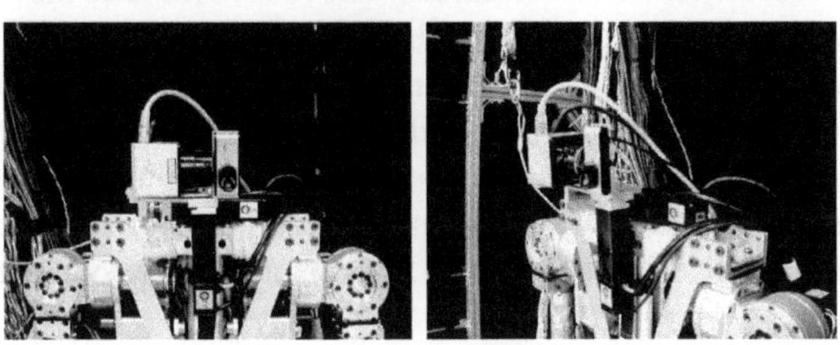

motors are pitch and yaw axes. In addition, a web camera is also installed next to the laser rangefinder.

2.5. Locomotion Mode

We model the robot as a 3D inverted pendulum, same as the one for bipedal walking (Kajita et al. 2003). The supporting point of the pendulum is assumed to be the point of contact. Then, only the heeling force *f* and the gravity act on COM. We use the crawl gait to show a quadruped walking (Aoyama et al. 2009). In this gait, the idling leg changes as follows: left rear leg, left front leg, right rear leg, and right front leg, in that order. It is designed so that three feet always touch the ground. The COM moves within the triangle which is formed by the three supporting feet. The transition from bipedal to quadruped posture is made by keeping static balance. Before transiting the posture between bipedal and quadrupedal stance, the robot stops walking.

3. Locomotion Stabilization

Locomotion stabilization is executed along the algorithm shown in Figure 6. As prediction for locomotion, the robot determines the parameters for locomotion mode, walking velocity, direction and numbers of paces. The robot has bipedal walking and quadrupedal walking as its locomotion modes, and on flat ground it travels in bipedal walk because flat ground is easy to negotiate. If on a slope or rough ground, it is impossible to move in a bipedal state, robots select quadrupedal walking since it is more robust. We propose recognition and planning using a laser rangefinder. A laser rangefinder enables a robot to recognize slopes or steps. Thus, the robot modifies the position of landing or COM position as it adapts to the environment.

As a feedback for locomotion, this stabilization scheme evaluates stability based on internal conditions. The robot estimates the risk of falling down using parameters which have uncertainty. If the risk of falling down is high, the robot changes walking velocity or direction, or selects internal models. But if the risk is still high, it changes locomotion modes. In this chapter, we will explain the method of estimating the risk of falling down using Bayesian Networks. In estimating it, we set "Robot Model Reliability (Reliability of Internal States)" and "Environmental Model

Figure 6
Locomotion Stabilization Scheme

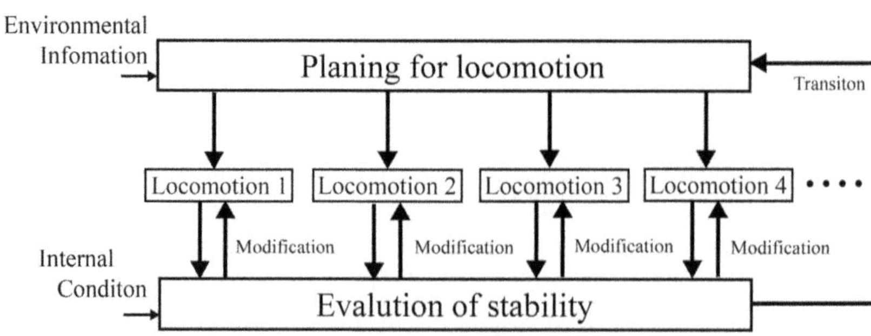

Reliability (Reliability of External Dynamics)." Reliability of a robot model shows how much difference there is between the motion in reality and the locomotion algorithm, or in short, the physical abilities of robot. For example, if the robot has motor trouble, the reliability of the robot is low and the risk of falling down is high. Reliability of an environmental model shows how accurately a robot recognizes environment. If the robot moves in dark, it does not get information of the environment; thus this parameter is low and the risk of falling down is high. In bipedal and quadrupedal walking, the robot evaluates both reliabilities, estimates the risk of falling down and attains an optimum gait adapting to the environment or the conditions.

4. Stabilization Based on External Information

We propose the method of calculating a gradient of ground which the robot faces to, and the boundary between two planes (between a flat and a slope, a flat and a wall) using a laser rangefinder. The gradient and the boundary determine how the robot transfers from a start to a goal. Figure 7 and Figure 8 are schematic diagrams showing that the robot measures an unknown ground. The laser runs in parallel with the sagittal plane. And the neck motor rotates around yaw axes, so the robot can acquire information about the 3D surface. Therefore, we define the system of Cartesian coordinates so that their origin is located at the point directly under the laser rangefinder and design a method to recognize a landform in the x-y-z 3D space. The robot needs to get (x_i, y_i, z_i) to estimate the landform. In Figure 7, a_i is the angle of laser i to the downward direction, and d_i is the distance between the center of the laser range-finder and the point on the ground surface. This data is obtained from the laser rangefinder.

Figure 7
Measuring Slope 1

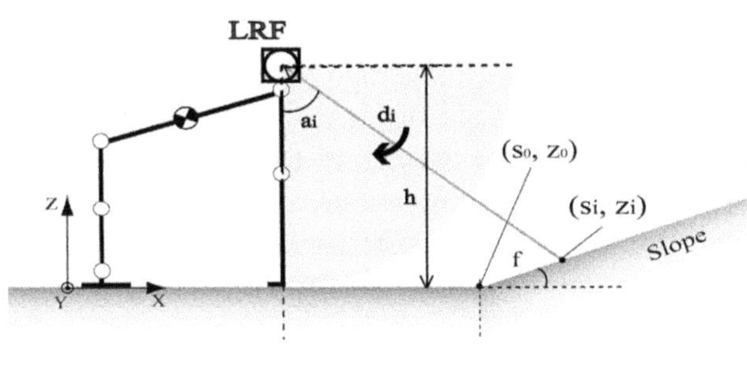

$$x_i = s_i \cos \beta$$
$$y_i = s_i \sin \beta$$
$$i = 1, 2, \ldots, m.$$

Figure 8
Measuring Slope 2

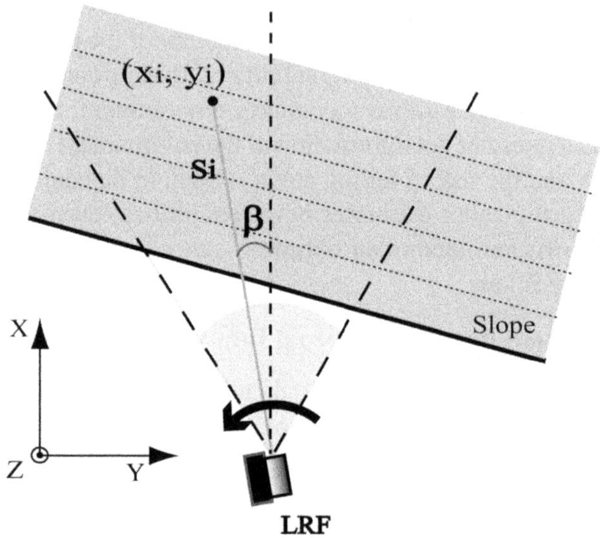

$$s_i = d_i \sin\alpha_i$$
$$z_i = h - d_i \cos\alpha_i$$
$$i = 1, 2, \ldots, m.$$

In addition, h is the height of the laser rangefinder. Then, a point on the ground surface can be described as (s_i, z_i). The coordinates x_i and y_i are calculated from s_i using the angle of the neck motor, β (Figure 8). β varies from -20 degrees to 20 degrees, every 5 degrees.

The equation of the ground surface is extracted by least-square method as $z = \alpha x + by + c$, where a, b and c are fixed numbers. Thus, the robot modifies the position of landing and the *COM* trajectory with this equation (Sawada et al. 2008).

5. Stabilization Based on Internal Conditions

5.1. Estimation of Probability

In this section we deal with uncertainty in locomotion to evaluate the stability. There are many kinds of uncertain parameters which have various dimensions; thus it is difficult to deal with them uniformly. These parameters are integrated into the risk of falling down as belief using the Bayesian Network. The Bayesian theory assumes that parameters have individual distributions, and posterior probability is induced formally by conditional probability. The Bayesian Network is a model

Figure 9
Bayesian Network for Locomotion

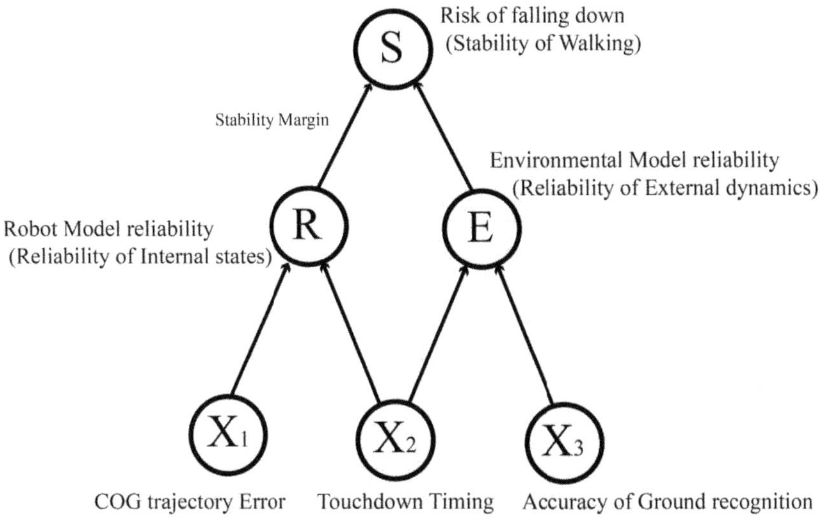

which describes relations among phenomena using probability. We describe the causality between the risk of falling down and the uncertainty parameters.

In our research, the Bayesian Network shown in Figure 9 is used to estimate the risk of falling down. First, the Bayesian Network estimates Robot Model Reliability "*R*" and Environmental Model Reliability "*E*." Reliability of a Robot Model *R* shows how ideal the robot motion is, and describes the capacity of moving. Reliability of an Environmental Model *E* is an index which shows how correctly the robot perceives the dynamics between the environment and itself. Secondly, the risk of falling down "*S*" is derived from *R* and *E*. "$S = 1$" shows falling down, and "$S = 0$" shows not falling down. Probability variables *R* and *E* have classes 0, 1, 2 in a more reliable order. Then conditional probability $P(S|R, E)$ reflects the performance of the robot, and the designer arranges this probability subjectively. The evaluating parameters X_1, X_2, X_3 shown below are observed in real time. Then probability variables range from 0 to 4, based on uncertainty which the parameters have input the Bayesian Network. When the probability variable is 0, the situation is most stable. The calculation of Bayesian Network uses the enumeration method shown below.

$$P(S=1)=\frac{\sum_{R=0}^{2}\sum_{E=0}^{2}P(S=1,R,E)}{\sum_{S=0}^{1}\sum_{R=0}^{2}\sum_{E=0}^{3}P(S,R,E)}$$

$$=\frac{\sum_{R=0}^{2}\sum_{E=0}^{2}P(S=1|R,E)P(R|X_1,X_2)P(E|X_2,X_3)}{\sum_{S=0}^{1}\sum_{R=0}^{2}\sum_{E=0}^{3}P(S|R,E)P(R|X_1,X_2)P(E|X_2,X_3)}$$

Since the evaluating parameters X_1, X_2, X_3 are always observed, each probability $P(X_1)$, $P(X_2)$, $P(X_3)$ is set to 1.

5.1.1. COM Trajectory Error X_1

The position of the center of gravity is measured by the force sensor which measures the force the robot exerts on its four legs. Outputs of sixth axis force sensors makes ZMP in bipedal posture. In quadrupedal posture, the center of gravity is calculated with the equilibrium of moments. Then, the errors between the desired trajectory and the observed trajectory decide the probability variable X_1.

5.1.2. Touchdown Timing X_2

The touchdown timing shows actual differences between the landing and the ground surface. If the robot is thrown off balance or the recognition is inadequate, the touchdown timing is earlier or later than the planned timing. The probability variable X_2 is renewed at every landing during walking.

5.1.3. Accuracy of Ground Recognition X_3

This parameter evaluates the performance of the recognition which the robot has. This shows how much information the robot attains with some sensors, and how abstracted is the environmental model which the robot has. Since the laser rangefinder has effective ranges, over this range there is much uncertainty. The two-dimension recognition and the approximate algorithm have the uncertainty.

5.2. Consideration of Stability Margin

The conditional probability $P(S|R, E)$ describes the influence which Reliability of a Robot Model R has with the Risk of Falling Down S. When the stability margin is large enough when compared with the COM errors, the influence is little, even if R goes down. In reverse, when the stability margin is small, R has a big influence on S. Therefore, $P(S|R, E)$ is decided based on the stability margin. For example, a stability margin in bipedal posture is smaller than one in quadrupedal posture, so $P(S|R, E)$ in bipedal posture is larger than in quadrupedal posture.

5.3. Shift of Locomotion Mode

The evaluating parameters X_1, X_2, X_3 are observed in real time, and the probability of falling down is estimated. The conditional probabilities used in the Bayesian Network are arranged by the subjective judgments of the designer. Therefore, the probability of falling down is not always 1 when the robot falls down. Thus, we pay attention to the fluctuation of the probability. That is, when the robot moves in bipedal posture and the risk of falling down increases, then it has the transition motion from bipedal to quadrupedal posture and goes quadrupedal walking. On the contrary, when the risk decreases in quadrupedal walking, the robot stands up and walks in bipedal posture.

6. Experiments

6.1. Experimental Conditions

In this experiment, the robot measures the landform with the laser rangefinder at starting point. During the walking, the robot decides the gait based on the risk of falling down estimated by the Bayesian Network shown in Figure 10. When the risk is more than β (0.7) in bipedal posture, the robot squats in quadrupedal posture, and when the risk is less than α (0.3) in quadruped posture, it stands up. The robot in bipedal posture has three patterns of bipedal walking $\alpha 1$, $\alpha 2$, $\alpha 3$ which have different levels of efficiency. If the risk decreases, the robot adopts a more efficient gait. In our research, this efficiency is the walking velocity. $\alpha 1$, $\alpha 2$, $\alpha 3$ are set as 8.67, 6.67, 4.67 cm/sec, respectively. The quadruped walking velocity is set as 3.00 cm/sec. Both the standing up motion and the squatting motion take 10.0 sec to activate. Modifications of the gait are conducted in every walking cycle. The robot aims to minimize the risk and maximize the efficiency all the time.

6.2. Experiment 1 (Planning Based on Recognition)

In this experiment, the robot walks from flat terrain up a slope. This slope inclines at 15 degrees and it is impossible for the robot to walk in

Figure 10
Velocity—Risk of Falling Down

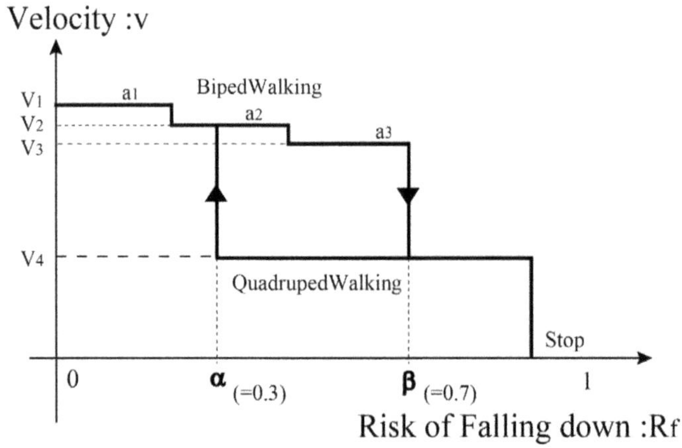

bipedal posture. At first, it estimates the slope information in bipedal posture at the starting point. Based on it, the numbers of steps in bipedal walking are decided. Then, the robot starts bipedal walk and stops it at the pre-determined steps to transit to quadrupedal posture. Finally, the robot climbs up the slope in quadrupedal posture. Figure 11 is the slope information acquired by the laser rangefinder. The gradient error is 1.5 degrees. Snapshots of Experiment 1 are shown in Figure 12.

Figure 11
Data for Recognition of Slope

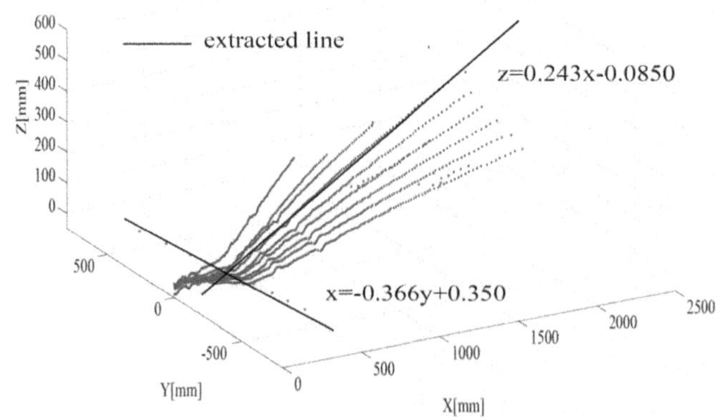

Figure 12
Snapshots of Experiment 1

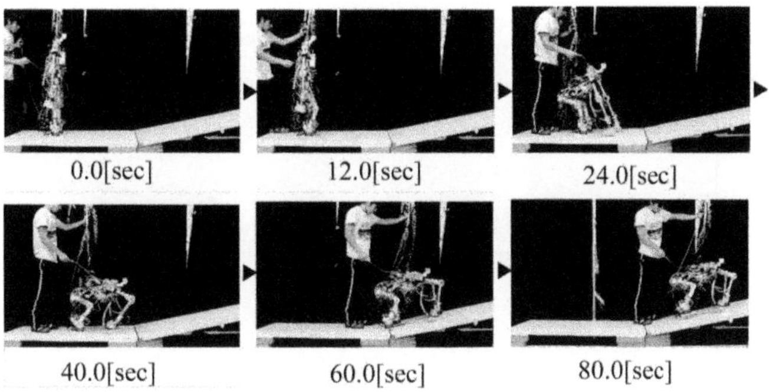

6.3. Experiment 2 (Transition Based on Risk)

In this experiment, the robot walks on rough terrain. There are inequalities which have the maximum height of 5 mm. This height cannot be recognized by the robot on purpose. We confirmed whether the robot in bipedal posture changes the gait to quadruped mode because the risk increases.

Figure 13 shows results about the COM trajectories derived from the force sensors. The COM trajectories induce X_1 shown in Figure 14. Figure 15 describes the probability variable X_2. The numbers in these

Figure 13
Comparison between Desired and Actual COM Trajectory

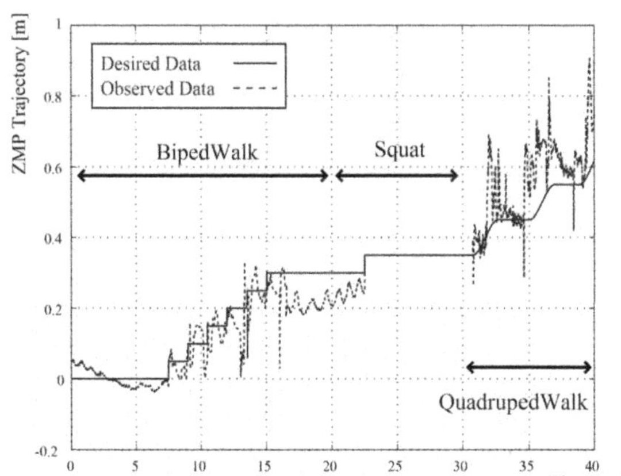

Figure 14
Experimental Data of Node X1

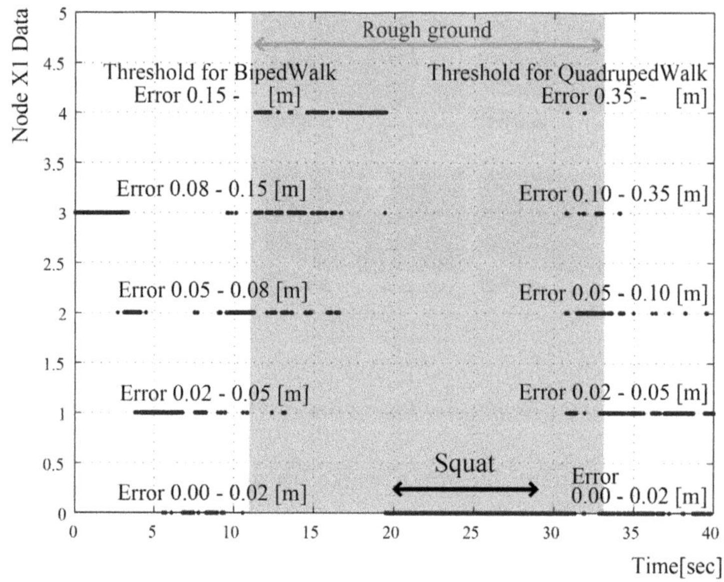

Figure 15
Experimental Data of Node X2

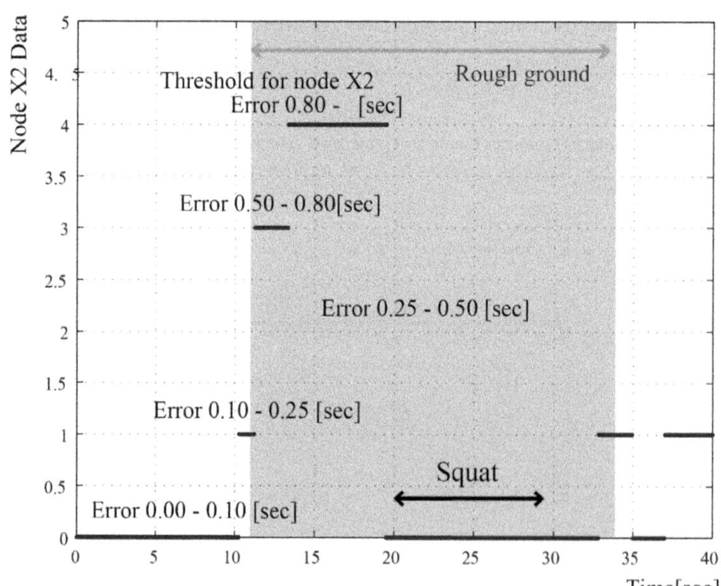

Figure 16
Risk of Falling Down

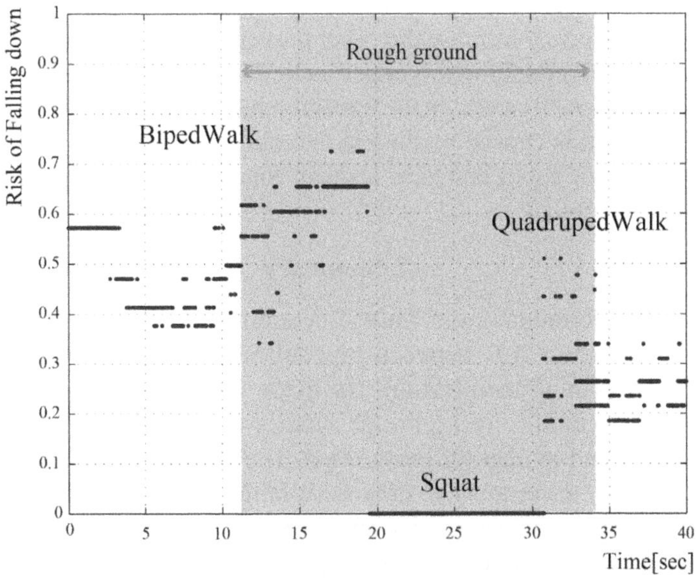

Figure 17
Snapshots of Experiment

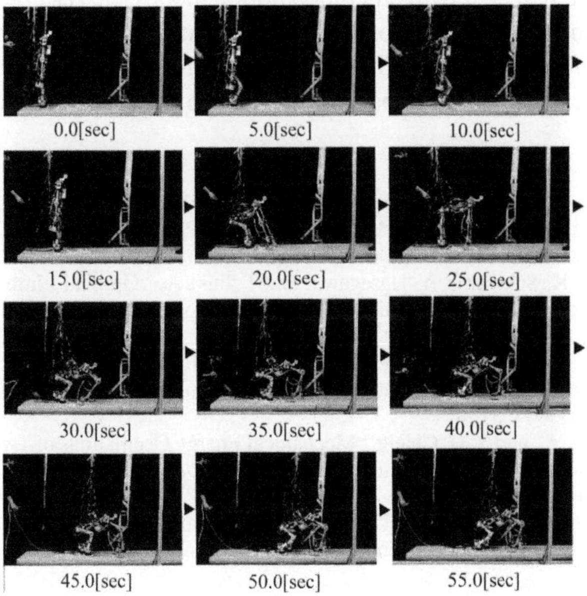

figures are the threshold to apportion the probability variable. In this experiment, the node X_1, X_2 have 0, 1, 2, 3, and 4 as the probability variables. When the probability variable is 4, the robot almost falls down. The node X_3 is always 0 because the robot moves within the effective ranges of the laser rangefinder in this experiment. Figure 16 is the risk estimated by the Bayesian Network. In the transition motion, the risk is 0.0. We can see the transition caused by the risk increasing. The risk is more than β (0.7) before the robot conducts a squat. Snapshots of Experiment 2 are shown in Figure 17.

7. Summary

We have introduced the Multi-Locomotion Robot (MLR). Motion selection of the MLR between bipedal and quadrupedal walk were described. In motion selection, we made two propositions. The first one was a method to recognize an unknown test course with the laser rangefinder and to plan for locomotion. The second one was the way of estimating the uncertainty as the risk of falling down. The robot decides the gait based on the recognition and the risk. We experimentally verified that the proposed methods can be applied to various cases, and showed that stable locomotion with transition between bipedal and quadrupedal walk has been accomplished.

Although we dealt only with bipedal walk and quadrupedal walk in this chapter, we will deal with other locomotion modes such as brachiation and ladder-climbing for transition later. Furthermore, we will add the Bayesian Network for the risk of falling down to more parameters, and diagnose the main causes which impart uncertainty to the robot system.

References

Aoyama, T., Y. Hasegawa, K. Sekiyama, and T. Fukuda. "Stabilizing and Direction Control of Efficient 3-D Biped Walking Based on PDAC." *IEEE/ASME Transactions on Mechatronics* 14, no. 6 (2009): 712–18.

Aoyama, T., K. Sekiyama, Y. Hasegawa, and T. Fukuda. "Optimal Limb Length Ratio of Quadruped Robot Minimizing Joint Torque on Slopes." *Applied Bionics and Biomechanics* 6, nos. 3–4 (2009): 259–68.

Beer, R. D., R. D. Quinn, H. J. Chiel, and R. E. Ritzmann. "Biologically Inspired Approaches to Robotics. What Can We Learn From Insects?" *Communications of the ACM* 40 (1997): 31–8.

Bertram, J. E. A., and Y. H. Chang. "Mechanical Energy Oscillations of Two Brachiation Gaits: Measurement and Simulation." *American Journal of Physical Anthropology* 115 (2001): 319–26.

Bertram, J. E. A., A. Ruina, C. E. Cannon, Y. H. Chang, and M. J. Coleman. "A Point-Mass Model of Gibbon Locomotion." *Journal of Experimental Biology* 202 (1999): 2609–17.

Chang, Y. H., J. E. A. Bertram, and E. V. Lee. "External Forces and Torques Generated by the Brachiating White-Handed Gibbon (Hylobates Lar)." *American Journal of Physical Anthropology* 113 (2000): 201–16.

Donelan, J. M., R. Kram, and A. D. Kuo. "Simultaneous Positive and Negative External Mechanical Work in Human Walking." *Journal of Biomechanics* 35 (2002): 117–24.

Eimerl, S., and I. DeVore. *The Primates*. Amsterdam: Time-Life Books, 1966.

Formal'sky, A., C. Chevallereau, and B. Perrin. "On Ballistic Walking Locomotion of a Quadruped." *International Journal of Robotics Research* 19, no. 8 (2000): 743–61.

Herr, H. M., and T. A. McMahon. "A Trotting Horse Model." *International Journal of Robotics Research* 19, no. 6 (2001a): 566–81.

———. "A Galloping Horse Model." *International Journal of Robotics Research* 20, no. 1 (2001b): 26–37.

Hoyt, D. F., and C. R. Taylor. "Gait and the Energetics of Locomotion in Horses." *Nature* 292, no. 16 (1981): 239–40.

Kajima, H., M. Doi, Y. Hasegawa, and T. Fukuda. "A Study on a Brachiation Controller for a Multi-Locomotion Robot—Realization of Smooth, Continuous Brachiation." *Advanced Robotics* 18, no. 10 (2004): 1025–38.

Kajita, S., F. Kanehiro, K. Kaneko, K. Fujiwara, K. Yokoi, and H. Hirukawa. "Biped Walking Pattern Generation by a Simple Three-Dimensional Inverted Pendulum Model." *Advanced Robotics* 17, no. 2 (2003): 131–47.

Lu, Z., T. Aoyama, K. Sekiyama, Y. Hasegawa, and T. Fukuda. "Transition Motion from Ladder Climbing to Brachiation with Optimal Load Allocation Control." *Advanced Robotics* 26, nos. 8–9 (2012): 1075–98.

Muybridge, E. *Animals in Motion*. New York: Dover Publications, 1957.

Saito, F., and T. Fukuda. "A First Result of the Brachiator III—A New Brachiation Robot Modeled on a Siamang." In *Artificial Life V*, 354–61. Cambridge, MA: MIT Press, 1997.

Sawada, H., K. Sekiyama, M. Kojo, T. Aoyama, Y. Hasegawa, and T. Fukuda. "Locomotion Stabilization with Transition between Biped and Quadruped Walk based on Recognition of Slope." In *Proceedings of the IEEE International Symposium on Micro-Nano Mechatronics and Human Scienc*e (2008): 424–29.

Simons, E. L. *Primate Evolution: An Introduction to Man's Place in Nature*. New York: Macmillan, 1972.

Maintenance Robot: A History and Reflections of a Japanese Engineering Scientist on Fukushima

Hiroyuki Yoshikawa
Center for Research and Development Strategy
Japan Science and Technology Agency
Tokyo, Japan

1. Preface

We learn from history of science and technology that results of basic research in science were applied to practical technology giving great benefits to society. It does not matter when the application appears, sometimes immediately or after more than a hundred years in other cases. When the age is for economic development, it is acceptable where any application is welcome by society. On the contrary, in the age of sustainability, some technology is longed for that urgently copes with problems disturbing sustainability, because many of those problems grow so quickly that the cost of solution increases rapidly as huge as not payable. So to speak, there appears a severe speed race between deterioration of sustainability and development of technology to suppress it. The developing engineers together with scientists are requested to understand this situation; recognizing the importance of discovery of such problems which they must tackle as soon as possible. This paper explains our experience in the past in which we, researchers, could not effectively give influences for national or company policies to prioritize the research and development of maintenance technology for improving safety, protecting people from the danger in failures of artefacts such as a nuclear power plant. This was exactly due to lack of communication between policy makers and researchers. I am now cooperating with policy makers to develop efficient paths of communication in Japan. The following paper deals with our failure of communication, but gives us significant knowledge for and a logical base of the present cooperation between scientists and policy makers.

In 1979, thirty-four years ago, my colleagues and I developed a prototype of maintenance robot to be applied to a nuclear power plant, not only for emergency but also for improving the safety of the plant and decreasing the exposure of maintenance workers to radiation. After completion of a practical robot in 1982, we brought it to a Japanese electric company, but unfortunately it was refused by the company. However, in the severe accident at Fukushima Nuclear Power Plant in 2011, we realized that a robot is a necessity there. There was, however, no robot available for this use in Japan. I regret it. Robotics research was quite popular in Japan already in 1970, but since then it followed two antipodal ways: one is for practical use in factories for welding or assembly and another is for theoretical research on complicated control of robots such as biped humanoid robots. Although maintenance or emergency robot we had aimed at could have been realized by integrating both technologies, neither basic researchers nor industrial engineers were interested in the integration and we became minority and could not get research money. I should confess that our effort was insufficient. More importantly, however, it was necessary for us researchers to develop "science of maintenance" generally which might enumerate the components of maintenance or emergency works and suggest the possible automation by robots among them. Science of maintenance is a tough field and less study has been devoted so far in the science and engineering community. In this paper, we shall look back at the experiences of our efforts to do it: including construction of theory, systematization of works, and realization of robot for maintenance. They are suffering and frustrating processes where researchers have fewer chances to write scientific papers. Present paper shows a short history of maintenance research mainly based on our own activities. After Fukushima, March 11, 2011, we observe some changes in the stream of development of robots in Japan from humanoid robot which pursues the fundamental behavior of human beings toward practical robots replacing human jobs which are severe and dangerous to humans. I believe that it is beneficial to look back to the history of maintenance research for creating the new stream of development of the robot in future.

2. Significance and Difficulty in Maintenance Technology

2.1. Prosperity and Maintenance

It has been more than two years since the nuclear accident at Fukushima which was caused by a big tsunami. We are still in the aftermath of the unprecedented occurrence. In 2012, many investigations were organized and four reports have been published: Fukushima Nuclear Accident Independent Investigation Commission of the National Diet, Investigation Committee of Japanese Government on the Accident at Fukushima Nuclear Power Station of TEPCO, Independent Investigation Commission on the Fukushima Daiichi Nuclear Accident (private sector), and Fukushima Nuclear Accident Investigation Report by Tokyo Electric Company. In every report, the process of escalation of the accident was scrutinized mainly in view of the activities of Cabinet, government, TEPCO, self-defense forces, and fire stations. The process of escalation of accident in physical and technical points of view, however, is reported only superficially. It is said that there was no necessary physical data available when those reports were compiled. It is quite understandable that the data were not available at those situations where almost all sensors were damaged and none could approach the site of accident due to high radiation. Although understandable, there is a difficult problem here that such situations will continue for unforeseeable years from now and still we, engineers, are obliged to learn lessons from the physical and technical aspects of the accident proposing new techniques for improving safety and no repetition of such a severe accident in future. We are definitely responsible and obliged to do it independently of the political decision of whether the nation will continue the nuclear energy or stop it. In Japan, teams of nuclear engineering specialists such as Atomic Energy Society in Japan have started scientific investigation of the accident and are expected to issue technical reports, but results are not yet available at this moment.

There should be, however, another aspect to learn lessons from Fukushima. It is a question on the direction of progress in technology which is backed by engineering sciences. The author has long been interested in establishing new engineering disciplines such as general theories of design, manufacture, maintenance, and so forth, which are to be devoted for actual activities in industries. The new disciplines should be practical but universally theoretical, applicable to any engineering field, which are different from traditional ones such as mechanical, electrical, material, bioengineering, nanoengineering, etc., which are based on scientific knowledge of specific domain and applicable to the particular technologies. The new discipline is also expected to be useful to develop universal equipment for various maintenance works. General theory and universal equipment complementarily work for establishing the particular discipline useful to improve the practical technology.

In 1970s, Japanese economy was progressing rapidly owing to the competitive manufacturing industries in many fields. People thought that it is better for mankind to supply more artefacts, such as electrical appliances, motorcars, and other commodities, together with comfortable houses, urban and transportation systems, and modern infrastructures. On the other hand, people knew the warning by Club of Rome that mankind will soon be faced by the scarcity of things: energy, food, material, and even space to live in, but they hesitated to think about this issue seriously within the progress of prosperity before them. In such a situation, however, there were many scientists even manufacturing scientists, who wished to tell people about the danger of such trends from a scientific viewpoint.

We had tried to develop engineering disciplines that would carry messages to people, because scientists should try to convince people by research to create theories practically applicable to change the attitude by answering the quest of Club of Rome, while satisfying the economic prosperity. We thought that maintenance of artefacts is one of the most important keys. We started struggling to develop a general maintenance theory which was a tough subject. The following is the record of our thoughts and practices in those days.

2.2. Warning the Necessity of Maintenance

In 1984, we presented a paper[1] on increasing necessity of maintenance according to the increasing “artefacts” in developed countries. At that time, however, the word “maintenance” was thought to embody nuances of underdevelopment or obscurity and gave unattractive impressions. In fact, the maintenance was not considered to be part of the main stream of industrial activities and was not viewed positively in everyday life. Even at the national level, the maintenance (or conservation) of the natural environment had drawn much public attention under the name of environmental protection, but its direction was still uncertain and confused. Even engineers, who recognize its significance in the industrial activity, may still feel it's not important enough to warrant their utmost efforts. In fact, it was rare for "maintenance specialists" to have a promising future in their companies and it was generally said among researchers that the productivity of papers for engineering studies on maintenance is low.

Surprisingly, we must repeat almost the same comments about maintenance even now. At present, when we read the preprint of our paper from 1984, we were astonished to know that the concept of maintenance has been kept almost unchanged for thirty years.

There are many reasons for the above. The profile of the maintenance problem is still vague and it is difficult to precisely define the effects of maintenance. Under such circumstances, it seems natural that maintenance problems are only discussed as local issues, even if an actual problem is very serious. This tendency has meant that the role of maintenance has not been adequately defined within the framework of technology as a whole. For example, various failures of artefacts such as aircraft accidents, railway accidents, wide area blackout, or fires in industrial complexes or urban high-rise buildings created serious debates in society after each accident about the safety of particular technologies applied. Natural disasters were also discussed in the same way because the damages caused by them mainly depended on technical systems involved in disasters. These debates were to claim the lack of management of gigantic technology, lack of man-machine systems, or the failure to predict and detect deterioration. These are all deeply related to the maintenance problem, but today only local or partial improvements result from these accidents and there is no discussion of the basic meaning of maintenance within the overall framework of technology. In the field of industrial or engineering research, research and development of new industrial products and new production techniques is much more attractive to engineers than maintenance. They are believed to provide great benefits, low-cost products, and a comfortable living environment to the people. Historically, the main objective of technology was to conquer severe natural environments and to replace them with comfortable technical artificial environments. It is reasonable, therefore, that engineers are more interested in developing new industrial products or production techniques. Although it was widely understood that maintenance recovers the functions of deteriorated products, and consequently it is beneficial and sometimes crucial to protect economy against loss, people, including even engineers, unfortunately did not feel particular attractiveness to maintenance.

The above attitudes are based on the understanding or thought that the object of technology is to replace nature with a comfortable artificial environment, which had originated from the early history of human beings who necessitated technologies for their survival. Technologies have not only been increasing the abundance of human beings for a long time but also protecting them from evils such as fierce animals, pestilence, germs, and natural disasters.

As stated above, the history of technology can be represented by the change from a natural environment to a comfortable artificial environment. In this respect, technology is still effective in developing countries. However, in many industrialized nations, artificial environments have been already full in space. Social capital stock such as roads and ports, many buildings in urban areas, various production facilities in manufacturing industry, and privately owned houses and consumer-durable goods are all our assets. These have been obtained by "improving" nature. However, our ability to assimilate these assets is limited. Firstly, natural land resources are limited and it will become difficult to find additional space for the new social capital stock and other artificial structures. It can also be argued that the benefits which can be accrued from the improvement of nature have been exhausted. Secondly, due to the limitations of resources and energy, it gradually becomes more expensive to scrap the existing stock to build a new one.

Moreover, recently in industrialized countries and perhaps soon in developing countries, artefacts which were developed for our prosperity will become new evils (modern evils in contrast to the past evils such as fierce animals) when they deteriorate and come to ages. As pointed above, the scales of natural disasters often depend on the size of technological systems involved. Especially, deteriorations of big industrial systems such as power plants and blast furnaces, and social infrastructures such as lifelines, highway systems, and buildings are urgent issues to address for developed countries.

Here, a new concept should be put forward, in which maintenance is considered to be a new object on which our existing technological potential should be concentrated. This concept will provide the basis for a philosophy of technology to properly position the maintenance problem within the entire scope of technology and then to cope in reality with this maintenance problem. We shall refer to some data obtained in the 1980s. There are two reasons to use such old data. One is that new data are not available currently because of the long-term neglect of economic importance of maintenance since 1980s until now. Another reason is that those data have considerable accuracy even now, enough to discuss the present problems. It should be stressed here that we have been lazy about utilizing those meaningful data which could be effectively applied to sustain the artificial environments for protecting people from dangers caused by their failures.

2.3. Corroborative Data

In response to the intuitive approach suggesting that maintenance will become more important in the developed countries, a question is immediately raised that economic activities may stagnate if maintenance is the only social activity. For example, what would happen to the present motorcar industry if cars could be kept in good condition and repaired to give a useful life of more than twenty years? Or, what would happen to unpaved roads which amount to 20 percent of all the roads in Japan if only maintenance of existing roads is performed without building new roads? If the shift from production to maintenance is performed intermittently and randomly, industrial and economic activities will be very badly affected by the discontinuity, resulting in a collapsing phenomenon.

The problem of shifting "from production to maintenance" should be considered as a concept of the changing trend in the objectives of technological efforts. There will be a trade-off between the production of new things and the maintenance of existing things. And the weight of maintenance in the industrialized nations is going to go on increasing, and this shift is occurring at a faster rate than the actual movements in various industries or changes in engineering research efforts. Therefore, objective or administrative measures are necessary to cope with this tendency, and hence it must be treated as a problem integral to the shift from production to maintenance. Examples of analyses based on this point of view will be explained below.

2.3.1. Social Capital

By viewing the formation of fixed capital by government such as roads, ports and harbors, airports, and other public facilities, it can be seen that there is the possibility of a sudden increase in the proportion of maintenance and repair costs for existing facilities to the amount of investment. On the one hand, social capital stock is generally increased by investment. However, on the other hand, there is an "amount of retirement" equivalent to depreciation over the life of a structure, which has no monetary value but is generally physically evident. Thus, value can be recovered by investing further capital. Although the required amount of this investment and the amount of recovery may not be accurately determined, they can be approximated by experience. Therefore, this amount of retirement can be considered as an index of potential maintenance and repair costs. On the basis of this approach, if a ratio of the amount of retirement to the formation of fixed capital in a certain year is called the retirement ratio, then this ratio can be used to express such intuitive quantities as the percentage of investment which is to be used for the maintenance and repair of existing facilities.

Values in the past and estimated values in future of the whole of social capital were determined from the concept described above and shown in Figure 1 below. (Lines are data until 1980 and estimation calculated for 1950~1980 based on the 1950 data. Some recent data of 2000 and 2005 and estimation of 2010, 2020, and 2030 based on 2005 data are available, shown by dots: Δ which are data from 2005 White Paper of Ministry of Land, Infrastructure and Transportation, Japan.)

According to this figure, the retirement ratio suddenly decreased from 1960 to 1965, during which the economic growth rate greatly increased in Japan, and it can be understood that the new investment performed a major role and the social capital stock was revitalized during this period. However, the direction of this tendency suddenly shifted after the oil crisis (estimated from 1980 data). Rate of increase (%) shown in this figure is the rate of increase in the formation of fixed capital by the government. If this rate is 0 percent, the retirement ratio increases to 37 percent in 2000 and continues to increase thereafter. If the rate is 6.5 percent, the retirement ratio becomes stable around 2000 and remains around 30 percent thereafter. Individually, roads and ports have seen similar progress, but facilities related to flood control have a high potential of suddenly increasing to more than 50 percent in 1980 and to 90 percent in 2000 (with a 0 percent

Figure 1
Percentage of Investment for Maintenance in Total Investment

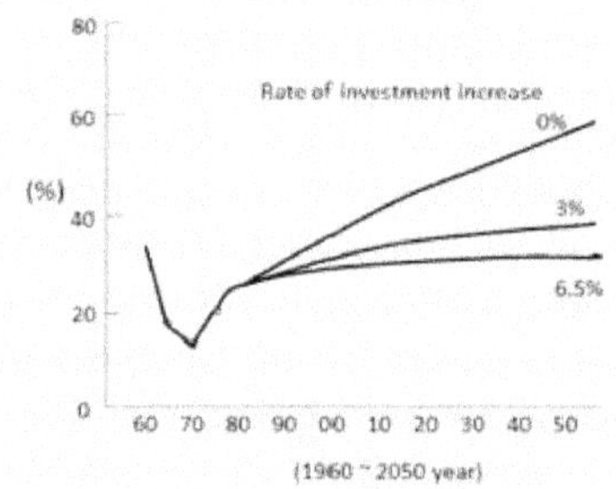

increase in capital formation). If the retirement ratio increases at a rate of 6.5 percent, then it stabilizes around 33 percent.

As seen in Figure 1, trends of recent data which are shown by dots are not so different from the 1980 estimation. That 2005 estimation of 2030 is higher than 1980 is the result of change of longevity due to the governmental revision of safety regulation of infrastructure after the Kobe earthquake in 1996. From this figure, it should be stressed that the estimation of calculated data in 1980 is useful. Unfortunately, it was ignored and no action was taken. And now, Japan is confronted with a threatening burden of a huge investment of maintenance of social capital, or infrastructure to protect people from danger caused by its failures.

2.3.2. Manufacturing Industries

It is difficult to discuss manufacturing industries systematically, as was done for social capital. This difficulty simply illuminates the core of the maintenance problem, because the approach to maintenance varies by industry and thus the evaluation of maintenance cost also varies and, in extreme cases, there is absolutely no concept of maintenance cost. Therefore, there is no unified way of estimating maintenance cost and an example of an estimate by Watanabe is shown in Table 1.[2] Maintenance cost varies by industry, but the steel and iron industry incurs the largest cost: amounting to about 5 percent of the total manufacturing cost. Total maintenance cost is about 2.4×10^{12} yen and the average maintenance cost is equivalent to 1.33 percent of sales revenue. This percentage is not increasing quickly at present but will increase considerably when the lives of facilities built during the high economic growth age end.

These uneven data in 1980, which look useless, imply significant facts. Many of the manufacturing industries still presently deal with the maintenance cost as a part of manufacturing cost. When economic or industrial activities are expanding, this kind of treatment is acceptable, for example, leaving the obsolete facilities aside. But when the activities become stable, their disposal or scrap and build of facilities becomes a significant part of the manufacturing cost. Unfortunately, many manufacturing industries in Japan have not paid enough attention to this even now in 2013, and consequently lost agility and recently put up bad fights in the international competition. The worst case is the neglect of waste of high cost radioactive material in Nuclear Power Industries. They left the used fuels on sites and considered no cost. Some economist is stressing that the manufacturing cost of electricity by nuclear fusion is far much higher than the present

Table 1
Maintenance Cost in Manufacturing Industry (Watanabe 1982)

	Amount of product (1000 million yen) A	Maintenance cost/ manufacturing cost (%) B	Manufacturing cost/ amount of sales C	Maintenance cost/ amount of sales (%) D=B×C	Maintenance cost (1000 million yen) E=A×D
Foods	20589	0.6	80	0.48	98.8
Pulp, paper	5615	2.1	80	1.68	94.3
Chemical	14968	3.5	76	2.66	398.1
Petroleum products	9698	1.71	88	1.50	145.5
Steel and iron	15773	5.0	80	4.00	630.9
Non-ferrous	6288	2.2	80	1.76	110.7
Metal products	9793	2.2	80	1.76	172.4
General machines	15590	0.7	80	0.56	87.3
Electric machines	18435	0.5	75	0.375	69.1
Transporting machines	21577	0.801	86	0.70	151.0
Precision machines	2972	0.5	76	0.36	10.7
Others	42959			1.22	481.1
Total	184257			Mean: 1.33	2449.9

one when considering the wasting cost of used fuels.[3] It is strange that the statistical data such as Table 1 of 1980 is not available at present. It is urgently requested to estimate them and be opened to public in order to establish the national policy of manufacturing industry.

2.3.3. Others

One industry showing high maintenance costs other than the manufacturing industry is the transportation industry with a cost of 4.8×10^{12} yen in 1979. In this sector, the automobile servicing share is 3.2×10^{12} yen, the railways share is 1.5×10^{12} yen, and the airlines share is 0.1×10^{12} yen, in addition, there is 0.8×10^{12} yen for electricity and gas, and 0.7×10^{12} yen for telecommunications.[2]

The data above show that the proportion of maintenance to overall economic activity in Japan has grown considerably. Though simplified calculations may give inaccurate results, the sum of 2.0×10^{12} yen for the social capital, 2.4×10^{12} yen for the manufacturing, and 6.7×10^{12} yen for the services will exceed 11×10^{12} yen indicating that maintenance is a huge sector, with 5 percent of Gross National Product (GNP) (1980). This value is also tending to increase every year and thus it is not appropriate to say that maintenance is a backward business.

However, other more important matters were observed behind the scenes. The "maintenance industry" in the 1980s was supported by labor-intensive activities in rather poor working environments as typified by road repairs, maintenance of unattended machine factories, periodic inspection of nuclear power plants, and maintenance of railway lines. It is surprising that nothing has happened in order to change this situation from the past thirty years to the present. Almost no progress of maintenance technologies such as mechanization and automation occurred. In addition, maintenance work requires multi-skilled workers with knowledge of many technical areas. This means that naturally there is a factor operating to concentrate wage increases within the maintenance industry. This also did not change in the last thirty years.

We have some evidences. The 1980s was the year of rapid rises in labor-productivity in manufacturing industries in Japan, which absorbed increase of wages and progressed competitiveness of Japanese products in the international economy. On the other hand, with respect to the maintenance cost, the production (amount of transport) per unit of maintenance cost dropped from 100 in 1975 to 90.5 in 1979 as shown in Table 2, so that rises in wages could not be absorbed earlier in the 1970s. When we observe the present status of maintenance in various industries, this drop

Table 2
Drop in Productivity of Maintenance

	1975	1976	1977	1978	1979
Passengers-kilometers + transported tons-kilometers ($\times 10^9$), (A)	109.3	109.6	113.4	116.1	118.6
Maintenance cost (10^9 yen)	162.7	183.9	204.5	215.3	233.1
Deflator (1975 = 100)	100	106.7	112.4	117.1	119.4
Maintenance cost ($\times 10^9$ yen) (price in 1975), (B)	162.7	172.4	181.9	183.9	195/2
(A)/(B)	0.672	0.636	0.623	0.631	0.608
(A)/(B) (1975 = 100)	100	92.7	93.9	93.9	90.5

in the productivity of maintenance is considered to have also occurred in other sectors and continued until recently, the 2010s.

2.4. Characteristics of Maintenance Work

Factory work was previously analyzed to automate machining factories and the following facts were clarified.[4] That is, work can be divided into several elements such as machining, installation and removal, transporting, measuring, and removing chips. These can be identified visually and thus are easily classified. However, if the functions of a machine deteriorate or a tool wears out earlier than expected, the worker changes the machining conditions or replaces the tool earlier, thereby eliminating the possible causes of a breakdown before it actually occurs. This is performed almost unconsciously by the workers themselves without being noticed by others and without recording. Therefore, if only the visually noticeable jobs are replaced with automatic machines, these unnoticed jobs will remain unchanged, thereby resulting in unexpected failures forcing systems to breakdown. Therefore, in order to secure a high rate of operation equal to that obtained before automation, it is necessary to employ a highly sophisticated "maintenance machine."

Generally, maintenance work requires the highly sophisticated or even intuitive judgments as well as skills of a human being. However, this is normally not clarified in the overall working system as described in the example above. This fact will make the calculation of maintenance costs more difficult especially in manufacturing industries and also make the clarification or defining of maintenance work technically more difficult.

In nuclear power plants, maintenance programs are examined in detail and the annual periodic inspection is regulated by law. However, according to the results of a survey[5] in 1982, the content of maintenance work for nuclear power plants is defined in the form of instructions to the workers, but the content would be insufficient if the maintenance were to be carried out by machines. This may be because the workers are capable of completing the work on the basis of the instructions, so additional information such as the "locus of motion of a worker's hand when closing a valve" is not necessary. However, in order to mechanize the work, information on the locus is vital which may be of a huge amount.

It can almost be argued that our knowledge of maintenance work is all at the level of the instructions given to workers. What's more, much of our knowledge is below this conscious level in machining factories and many unnoticed jobs are being performed almost unconsciously. Nuclear power plants, railways, and the aircraft industry can be considered advanced maintenance fields in which maintenance specialists are employed. In this case, the maintenance work is most likely to be specified in the form of instructions, but, still far from the necessary information for automation.

If we seek the automation of maintenance, we must clarify first what the maintenance work is. In 1980s, we have insisted the importance of maintenance automation in future and accordingly several surveys were conducted. Results were interesting and inspiring, and we used those results for designing a maintenance robot to be described in the next chapter. Some results are outlined below.

2.4.1. Interdisciplinary Character

It has been made clear that maintenance technology is not a province of a specific technical field or an industrial sector, but rather should be used in all technical and industrial fields. Although the needs of maintenance will vary qualitatively as stated before, maintenance embodies common phenomenological patterns of problems in all fields, such as occurrence of trouble, diagnosis, and repair; therefore, there is a strong need for the establishment of a general theory or universal technology of maintenance capable of handling all fields transversely. However, according to the results of a survey conducted by the author, highly sophisticated maintenance techniques have already been developed through large amounts of investment in almost every field, but these have been developed merely as techniques linked closely to each field and also there are almost no universal techniques or technologies transferable among different fields. Generally speaking, as the objects of technology become more diversified, the more difficult problems occur in formulating and using models, analysis methods, system composition methods, and optimization techniques for handling maintenance techniques with a common approach. In view of this, maintenance technology is considered to be one area in which systematization is extremely difficult to perform.

2.4.2. Diversity

Even if maintenance technology is to be developed only for a particular industrial field, there are still many difficulties to overcome. One of these is diversity. Let us consider a system failure with a system shutoff. To cope with this failure, the knowledge required to be applied is very diverse and is not limited to unique engineering domain such as mechanical or electrical, because the system components generally have the diversified attributes: electricity, fluids, chemistry, and so forth. Systematic theories based on this diversity will differ from conventional engineering theories. It may be possible even now to make an abstract theory by not considering the physical personalities inherent in these system components to simplify the problem, but it is too abstract and there is a lack of reality for the theory to be applied practically. In this case, the results obtained from the model will inevitably be arbitrary. True theory of real maintenance must contain even the modeling of systems, but the possibility of such a true theory will remain hereafter as a future problem.

2.4.3. Obscurity

Maintenance technology can be more accurately characterized by obscurity. If a failure occurs, the trouble spot will be identified and isolated by diagnosis and then corrected by repair. A series of techniques for this restoration work can be considered in terms of control theory. Control is categorized according to the level of knowledge about systems. Then, this control technology, namely maintenance, can be positioned as shown in Table 3.

From Table 3, it can be seen that diagnosis and repair deal with the unknown, and thus a new element in the detection of objects, which was not used in conventional control, is involved in diagnosis and repair. This makes the problem more complicated and very difficult to solve.

In view of the above, diagnosis and repair must be positioned as a very sophisticated subject, which was gradually complicated through

Table 3
Difficulty of Maintenance due to Many Unknown Factors

	Controlled spot in system	System characteristics	Controlled variables
Simple control	Known	Known	Known
Feedback control	Known	Known	Unknown
Adaptive control	Known	Unknown	Unknown
Diagnosis and repair	Unknown	Unknown	Unknown

the historical development of control technology. Theoretically, control technology has been developed up to the stage of adaptive control, but there are no theories for systematizing diagnosis and repair. This kind of theory must be developed in the years to come. Thus, it is considered that the present maintenance technology in each field is based on the ad hoc concepts predetermined. The difficulty of technology transfer between different fields is considered to occur mainly because of the lack of a "diagnosis and repair theory," which would form the common basis for different maintenance techniques. We need to formulate this theory as quickly as possible.

2.4.4. Non-Repeatability

Non-repeatability is another unusual characteristic of maintenance technology. Systems where failures occur frequently and regularly are revised by redesigning fallible parts and consequently fewer failures will occur thereafter. These fewer failures will occur without obeying any rules, so it is essentially difficult to find out a useful rule by which the maintenance policy would be derived. Therefore, the method based on experience and induction conventionally employed in various fields of engineering cannot be applied. On the contrary, it is more effective to first make hypotheses or assumptions and then to derive techniques for solving the problems that have arisen. However, for effectively utilizing this method with a high probability of success, there are still many theoretical difficulties and many problems that have to be solved in future.

These above-mentioned difficulties, 2.4.1–2.4.4, were already proposed thirty years ago, and some needs were strongly appealed and urgent actions were proposed: such as collection of maintenance data from wider areas of industry, communication and exchange of experiences among different industries, conducting research on the relationship between failure physics at universities and system failures in sites, and development of theory of maintenance which will be universally applied to practical problems. Only a few activities, however, were initiated at some universities in Japan. The reason was clear. It was that opening failure data to the public by firms even for scientific research was difficult because the failures of products and/or facilities were things that should not exist in the firms, and therefore they were reluctant to open those failure data even for the scientific research. Unfortunately, a research committee organized by the author, supported by government around 1990, could collect only limited data from very limited fields of industry. But the university members of the committee were not successful in developing useful theory.

3. Maintenance Robot for a Nuclear Power Plant

3.1. Enthusiastic Creation of a Robot for a Nuclear Power Plant

We have studied the design of maintenance robots which is to be applied to diversified maintenance jobs, namely universal maintenance robots. Such design studies need transdisciplinary consideration which resembles that of seeking general theory of maintenance. Besides the basic study of design, we constructed a maintenance robot for a nuclear power plant not only as a case study but also aiming at practical contribution to benefit maintenance workers in normal conditions to decrease the exposure to radiation and in emergencies to offer a crucial means to cope with difficult jobs in accidents of plants. We have made a prototype in 1978, one year prior to the Accident of Three Mile Island Nuclear Power Plant.

The author organized a research and development group in the Precision Engineering Department of Faculty of Engineering, University of Tokyo in 1978 with five young (at that time) professors of the department. We remember intensive and enthusiastic discussions among members for conceptualization of a maintenance robot to be practically applied to nuclear power plants. It was aimed to design a robot that moves freely, along narrow corridors and up and down the stairs in the containment vessel of nuclear plants, for observing equipment, diagnosing failures, and repairing the failures, by the aid of remote observation through camera and remote operation of flexible manipulator.

The research project was named "MOOTY" (later called AMOOTY), under the collaboration of young (at that time) professors who were engaged in different technological disciplines: material science, computer science, precision metrology, mechanics of precision machinery, robotics, and design theory in the Precision Engineering Department. The name MOOTY was the sequence of initials of the five professors' names. (Miyoshi, Odawara, Oozono, Takano, and Yoshikawa (chair). The first letter A of the later name AMOOTY stands for Arai who participated later.) The author believed in the success of the project, but it was not certain if it would be accepted by the electric power company which was responsible for the operation of the plants. The MOOTY was perhaps moot at that time.

The prototype is shown in Figure 2. The characteristic features of the robot are as follows:

3.1.1. Link and Wheel System for Move[6,7]

The wheel is special for the free movement in corridors and stairways in the containment vessel. It has a triangular link equipped with three small wheels on the tops. When it moves on the flat floor, the small wheels work and when climbing up and down on stairways the triangular links rotate to secure the motion along the steps. A mock stairway was made in the containment vessel at the university laboratory, which was an exact replica of the real one. It moved autonomously with the aid of laser type position detector identifying stairways and avoiding obstacles. The tests of MOOTY's move were quite successful.

3.1.2. Metamorphic 9-DoF Manipulator[8,9]

We obtained some basic requirements for manipulator as mentioned earlier. The first step of the conceptual design is to determine the geometric

Figure 2
Maintenance Robot

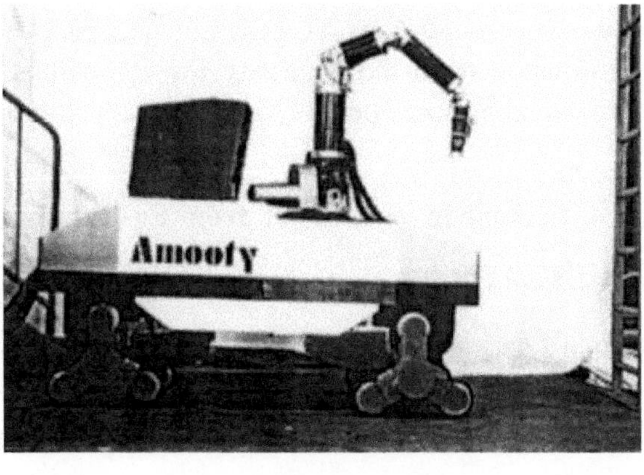

structure of the manipulator. The geometric structure from kinematic viewpoint denotes joint types, their sequence, dimensions, and offset directions of links. We have concluded to give more flexibility to it, namely 9-DoF (degrees of freedom of motion). This manipulator can also be regarded as a metamorphic one which can transform itself to eighty-four kinds of 6-DoF manipulators. These metamorphoses are classified into eighteen kinds of joint configurations. And in the same joint configuration, different link proportion structures are also available. We defined the metamorphic manipulator as follows: "A manipulator which can transform itself to another in different geometric structure."

Each of the metamorphoses has specific features respectively, typically on workspace. The possibility to accomplish the given tasks is judged by confirming whether the whole trajectory is included in the workspace or not. Taking an example of closing valves, it was revealed that even the location of the valves influences the structure of manipulators.

3.2. Refusal of Real Use

MOOTY for practical use was constructed by a company which is a manufacturer of nuclear power plants (Toshiba Co). We brought it to an electric power company asking maintenance specialists of the company to implement it practically. However, our proposal was refused. The reason of the refusal was that it is dangerous to install such a mobile device to containment vessels. It was understandable. But the second answer was the one which awfully discouraged us. They said that the company cannot support the MOOTY project because such a project might imply the potential danger of nuclear power. They were in the belief that their nuclear power plants would never fail so seriously that they need such a mobile maintenance robot. Their belief had no scientific evidence but was very tough. Now it is generally understood that this tough belief was originated from the necessity for companies to avoid the resistance movement by some people against the construction of nuclear plants.

4. Lessons to be Learned from Fukushima

4.1. Fukushima Accident

As mentioned earlier, the mitigating process at the acute phase of the Fukushima Accident was not systematic enough. It is said that the Prime Minister, the government, related ministries, the Nuclear and Industrial

Safety Agency, Nuclear Safety Commission, Self-defense Forces, Fire and Disaster Management Agency, TEPCO, related companies, government advisors, and foreign experts, etc. worked intensively to mitigate the disaster. They worked very hard individually but their work was not efficient because of insufficient organization to be formed at any emergency. We had no plan to launch a cooperative team integrating people of normally separate organizations in emergencies, especially in such disasters as Fukushima. This was the reason why the reports of accident elaborated only human behaviors. We think, however, that such an insufficient organization for prevention and mitigation of the accident at the acute phase comes from the lack of right and enough knowledge about the technological structure of the plant and of the method of diagnosis and repair based on them. This means that the responsibility of Fukushima should not only be taken by those who worked around the disaster but also by people who designed, manufactured, maintained, and managed, namely scientists and engineers of nuclear technology, and further, those of fields including wider areas of science and engineering, who are related to energy research. Social scientists and engineers are not exceptional. The author who made a maintenance robot and tried to install it into nuclear power plant has had exactly this responsibility. Moreover, he cannot excuse his failure in actual installation of MOOTY, because it is due to the negligence of scientist's intrinsic duties. Watching US robots working efficiently to observe the situation of failure in Fukushima plants where none could approach, the author deeply felt the responsibility.

We may say that Fukushima is now in the chronic phase. Sufferers must work hard and many people outside damaged area work to encourage and help the recovery of the area. But damage expanded widely, and recovery is slow. Many sufferers are still forced to live in temporary houses. We now recognize that again various kinds of knowledge in science and engineering are essential to promote the recovery, especially recovery into a state better than before the disaster.

4.2. From 1980 to Present: Responsibility of Scientists and Engineers

Here, we shall look back at what we have done in the area of engineering related to the nuclear failure, after our trial in 1980s. As mentioned in the preceding chapters in detail, we had exactly noticed the possibility of occurrence where machines such as maintenance robot would be inevitable.

4.2.1. Basic Research on Maintenance

There are two categories of research. One is physics of failure where various phenomena are studied that cause deteriorations and damages of the functionality of systems. Another is basic research on systemic theory of maintenance technology. The first one has considerable progresses: clarification of failure mechanism of material such as failure mechanics, fatigue theory, wear theory, erosion/corrosion theory, etc. including computational physics. On the contrary, the second has been poor. The author made a semi-quantitative theory of maintenance system, but unfortunately failed to apply it practically due to lack of real data of system failures on sites. Consequently, the gap between two categories has not been eliminated and thus a holistic view on the maintenance of systems has almost remained at the level of 1980s.

4.2.2. Maintenance Business

The states of maintenance in industry have not been essentially changed since 1980s. Techniques have been developed in some areas such as airline and railway business, and manufacturing industries such as steel making and chemical engineering. However, they are separate and hence no communications happen among them for exchange of technologies. Further, maintenance of infrastructures such as roads, highways, rivers, and forests has been insufficient. Consequently, exactly as was pointed out by our reports in 1980s, potential cost for maintenance is now estimated unusually high. Relevant policies, together with the development of maintenance technologies based on research, are urgently requested.

4.2.3. Research on Robotics

After Three Mile Island Accident, Japanese government implemented a national R&D project in 1983, which aimed at developing robots that work in an unusual environment called "Advanced Robot for Hazardous Environment." Several robots were developed for many purposes such as rescue robot, underwater robot, fire robot, etc. They were only prototypes and were not actually used as completed systems. Unfortunately, there was nothing for nuclear plants. After JCO accident at Tokai, Japan, 1999, government again asked nuclear plant makers to develop maintenance robot but did not actually use them in plants of electric power companies. Those projects were managed by Ministry of International Trade and Industry and university researchers were not involved; hence, no basic research took place on maintenance robots. Since the 1980s, researchers in universities had been very much interested in mobile robots. The main interest was the moves at flat floors of factory and laboratory. Flexible wheels were one of the most interesting subjects. Especially in Japan, after the widespread use of practical industrial robots at various operations in industries: such as transporting, welding, painting, and assembling, and moreover no more support for practical robots as maintenance robot from government, Japanese university researchers went into a more challenging field of robotics, namely developing humanoid, especially biped robots of human size (The first biped robot was WABOT by I. Kato, Japan, 1969). Extensive researches on the mechanism of walking or running styles of human beings have been conducted in order to acquire necessary knowledge for the development of mobile robots. Many of the researchers were interested in basic or theoretical aspects of robotics: how similar robots can be to the humans. The international level of Japanese robot technology together with the basic research on those topics has grown considerably high, but they could not work in containment vessels at Fukushima.

There may be various reasons for those problems. However, the author believes that scientists and engineers are responsible at least for the crucial factor: offering basic knowledge of maintenance which is one of the most important ones to establish sustainability in modern society.

5. Proposals

As an engineering scientist who worked in research on establishing general theories of design and maintenance for contributing toward the construction of sustainable society, Fukushima gave me a serious opportunity to reflect the habits of Japanese scientific and technological community.[10] We are confronted with crucial delay of relevant knowledge application against the rapid increase of artificial systems in society. Simply put, there was plenty of useful knowledge among scientific and technological communities but no use by society. Fukushima was a symbolic event. We need change.

First, scientists and engineers must remove barriers among different disciplines within the scientific community and industrial fields.[11] Unfortunately, there is no unique scientific or engineering field where scientists and engineers can discuss about the science and technology of maintenance for safety, security, and economy of artificial systems generally. Each of these subjects is studied in individual scientific domain and for specific systems. Therefore, good communication cannot exist between specialists. Under such circumstance, it is difficult to write a good scenario for developing sustainable technical systems. The accident of Fukushima which was the exact opposite of sustainability, should have been mitigated by continuous infusion of knowledge for system safety from other fields, which have remarkably developed in these decades through science and engineering studies and accumulation of practical experiences of industrial activities.[12]

Secondly, the author proposes to construct a better communication path between scientists/engineers and society. It is urgently requested that scientists and engineers should devote themselves more by integrating their knowledge and give independent advices to the society, including governments and policy makers who are asked to be more and more knowledgeable about science and technology when legislating and making national policies. We recognized that the regulation of nuclear safety has not effectively worked for improving the safety of plants in these decades. Electric power companies were only requested strictly to keep the regulation of maintenance and thus, have lost chances to improve the safety of plants implemented in the past. Policy makers should always consult with specialists in order to adapt the regulation to the frontier of scientific and engineering knowledge beneficial to improve the safety. I think that specialists, namely nuclear scientists and engineers are definitely responsible to lead the realization of relevant communication with the aid of scientific community.

The third is closely connected to the second, but concerned with the mindset of scientists and engineers for the better communication between specialists and society.[13] Because of the severe competition of acquiring research fund, researchers recently tend to concentrate their interest and effort into a very narrow subject which promises the winning. These researchers normally insist the importance of their own research subjects and petition the policy makers for priority. This is not a neutral or independent advice which is useful for policy makers to plan policies which are really requested by society. Therefore, researchers—scientists and engineers—should be able to propose neutral advices when requested. This capacity is possessed by scientists and engineers with broad knowledge of science and technology including their implications to society. I roughly estimate that scientists with broad knowledge and with intention to serve neutral advice are unfortunately less than 1 percent of all in Japan, at least apparently or actually influencing. Others only seem engrossed in their own narrow research subjects. It should be stressed, however, that this percentage is the result of the above mentioned competitive pressure on researchers, and therefore it can be changed by eliminating the pressure. The elimination is realized by admiring the scientist's neutral advices and giving them opportunities to get socially good positions in governmental and private organizations. This requests the change of mindsets of persons of influence and enactment of good regulations which hopefully wake up the young scientists with potential intention. There is an urgent need in Japan to provide an official framework in which we may find out and encourage the hidden researchers or university professors to act overtly as neutral advisors for science and technology. We are now working for it.

Notes

1. H. Yoshikawa, *Necessity and Potential of Maintenance Technology*, 7th E.F.N.M.S Congress, 1984, Venice.
2. Y. Watanabe, "Maintenance as an Economic Sector," *Journal of Precision Engineering Society* 48 (1982): 11 (in Japanese).
3. Kenichi Oshima, "Cost of Nuclear Power," Iwanami, 2012 (in Japanese).
4. H. Yoshikawa, "Manufacturing Automation and Human," *Journal of Mechanical engineers of Japan* 79 (1976): 692.
5. *Survey Report on Support Systems for Nuclear Plants*, Ministry of International Trade and Industry, 1982 (in Japanese).
6. M. Takano, G. Odawara, "Development of New Type of Mobile Robot TO-ROVER," 13th ISIR, Chicago, IL, April 1984. 20.81–20.89.
7. T. Arai, et al, "A Stair-Climbing Robot for Maintenance; 'AMOOTY'," *Seminar on Remote Handling Equipment for Nuclear Fuel Cycle Facilities*, International Atomic Energy Agency (IAEA), OECD Nuclear Energy Agency, Harwell, UK, October 1984.
8. H. Asama, and H. Yoshikawa, "Development of Metamorphic Manipulator with 9 Degrees of Freedom," I.S.R.I., 1985, Tokyo.
9. T. Miyoshi, and K. Arima, "Rigidity Evaluation of Articulated Robot Arms," (in Japanese) Proceedings of the Japan Society of Precision Engineering Autumn Session, Tokyo, May 1981, 823–26.
10. H. Yoshikawa, "The Role of Scientists in the Response to the Fukushima Nuclear Power Plant Disaster," Center for Research and Development Strategy, JST, April 28, 2011.
11. H. Yoshikawa, "Design Methodology for Research and Development Strategy," Center for Research and Development Strategy, JST, February 2012.
12. H. Yoshikawa, "Reformation of Scientific Disciplines," World Conference on Science, ICSU/UNESCO, Budapest, 1999, 76.
13. H. Yoshikawa, "New Roles of a Scientist," Iwanami - Shoten, 2002 (in Japanese).

Technological Singularity: A Path towards Artificial Intellects and Unpredictable Futures

Jack Brzezinski
McHenry County College, Canada

1. Abstract

Extrapolating current data about technological progress ten to twenty years into the future leads to technologies that many scientists and engineers consider implausible. However, a far-reaching upgradation of our bodies is already under way with biotechnologies and nanotechnologies merged with several fields of computing. Just as capabilities produced by artificial intelligence (AI) were inconceivable several decades ago, the new research opens new radical applications. Many scientists reject the exponential path to super advanced technologies or are simply too busy to reflect on the scientific big picture. Others embrace the seemingly far-fetched consequences and predict technological singularity—a period of extremely rapid technological expansion leading to the creation of super-intelligent artificial intellects. The key technological drives toward AI will likely be advances in neuroscience, genetics, nanotechnologies, and robotics. AI revolution affecting all aspects of our life is likely to be the most profound change that this planet will experience. In this context, the transhumanist characteristics of the research are an important reason for the skepticism and discomfort in the scientific community. No matter what the time frame for singularity turns out to be, it is only a question of time before humanity has to choose whether to embrace the progress discussed here. The scientific community, governments, and the popular media are increasingly interested in the impact of technological breakthroughs. Assuming that scientists and entrepreneurs can deliver their breakthrough projects within a couple of decades, we have a responsibility to engage in the discussion about implications of the research.

2. Technological Singularity: Skeptical Outlook and Great Opportunities

Discussions about the future technologies are increasingly focusing on topics regarded by many scientists and journalists as implausible. The technology in question is the general AI. In this chapter we will try to describe some of the key research fields and applications.

Relatively few people appreciate the deep implications of the exponential pace of the technological progress postulated in 1965 by a thirty-six-year-old research physicist at Fairchild Semiconductor Corporation by the name of Gordon Moore (2006). Moore predicted that the number of transistors in a chip would double every year. In 1975, he revised the doubling period to two years estimating that this trend will continue at most a decade longer. Today, microprocessors entering the market still follow the same, almost fifty-year-old trend (Stackhouse et al. 2009; Gelsinger 2006). Ray Kurzweil, a brilliant scientist who is one of the strongest voices predicting singularity, points out that we are witnessing the beginning of the sixth exponential paradigm. The key technology enabling the sixth paradigm is molecular level computing in three dimensions. The previous five paradigms in computational system were based on the following technologies: electromechanical, relay, vacuum tube, transistor, integrated circuit (Kurzweil 2005). If Moore's Law continues unstopped, around the year 2020 it will be possible to store one bit of information on a single atom (Seth 2002). Exponential models have been at work in many other areas of technology, for example, genetic research, shrinkage in size of mechanical devices and nanotech related patents, memory storage capacity (Koh and Magee 2008). Although statistical variability between fields has been documented, the overall trends are difficult to refute.

The arguments against technologies predicted to arrive in about ten or twenty years are simple and obvious. Many research questions are still unanswered. In the past, scientists were able to deliver answers but how long will they be able to continue doing so? Are the paths currently explored the right ones? In general, we can identify two main sentiments.

The first source of the skepticism is the belief that the current trends of exponential growth in key technological areas are about to end. The

Moore's Law, for example, has been a model accurately predicting the complexity of microprocessor designs for many decades. During that time, significant technological problems have been resolved. These hurdles did not affect the overall pace of growth. Currently, we are approaching another barrier leading to uncertainty about the future. New projects will have to take into consideration quantum effects. Similar uncertainty exists in many other sciences, which have been rapidly growing over the past decades. This point of view leads to the following conclusion: the super-advanced technologies might eventually arrive in a very distant future.

Secondly, the new technologies are considered implausible because they will be too complex. The technology in the center of this discussion is AI or more precisely the development of a super-intelligent artificial intellect. The field of AI became a mature discipline and has produced fascinating results. The bulk of the discussion in media, however, seems to be concentrating on whether computers can have "real" emotions or feelings. This is certainly an interesting philosophical question, which might even lead to opening new engineering avenues. However, some of the key capabilities of intelligent systems related to natural language understanding, machine translation, or image processing are rapidly evolving independently from philosophical discussions.

On the other hand, we can assume that the rate of progress remains approximately the same. In this case, the progress leads directly to the notion of technological singularity, which perhaps is even more contentious than the pace of progress or the complexity question. Singularity is a mathematical term which describes the breakdown of a model (Arnol'd 1981; Bruce and Giblin 1992). In robotics, the singularity definition has a bit different dimension, one that is very well-suited for our discussion. Singularity describes a robot configuration that cannot be controlled. There may be infinite forces acting within the robotic system (Merlet 2007).

Technological singularity, which is the focus of this discussion, is a term that describes the culmination of exponential pace of technological advancement and research in several areas of science. These areas are most likely to include computing, robotics, neuroscience, genetics, and nanotechnology although it is very possible that new discoveries will spawn new fields. Any exponential model follows a path that at a certain point becomes vertical. Beyond this point, the model designates practically an infinitely fast rate of increasing values. Technological singularity will be a result of exponential models at work in areas that deepen our understanding of the key domains of AI as we understand it today: neuroscience, machine learning, logic, knowledge representation, algorithms for natural language processing, and problem solving. Once we have an understanding of "the software and hardware of intelligence," the increasingly powerful computing architectures will be the platform implementing artificial intellects with general problem solving capabilities far greater than those of human brains. This possibility is perceived by many researchers and analysts as implausible. Even more implausible is the logical consequences of truly intelligent computing systems with problem solving capabilities. As the time passes, supporters of technological singularity believe that these systems will be self-improving autonomously. Non-biological computational systems or biologically inspired systems will reach complexity necessary to exhibit very high level of intelligence capable of solving any problem.

The singularity will probably happen without much public recognition and fanfare. Intelligent systems are likely to appear inside military supercomputing centers, search engine systems equipped with huge data storage and computing systems, commercial cloud computing server farms, or narrowly specialized consumer products such as autonomous car guiding systems or smart phone applications supported by large data centers. These systems will obtain capabilities that are not foreseen by engineers or programmers. In fact, there is a significant government funding available for supporting research into self-evolving and self-healing systems which are supposed to be immune to software or hardware failure.

Particularly in the academia, there is a resistance toward accepting bizarre results of extrapolating the current rate of progress into a couple of decades in the future. This perception is not very surprising. Scientists working on cutting-edge technologies are much more likely writing software or trying to overcome some technical difficulties rather than engaging in long-term predictions. Creation of artificial intellects and the merger of our biological bodies with the technology seems to be a particularly uncomfortable subject despite the recent spectacular successes of DeepQA in the Jeopardy TV show (Ferrucci 2010). DeepQA's ability to disambiguate highly nuanced natural language expressions and to generate coherent short answers is truly remarkable. Equally astonishing is the size of database processed by DeepQA to generate an answer based on sophisticated search and summarization algorithms. One can only imagine what kind of impact such a system might have when it becomes commercially available and applied in home electronics or handheld devices. We are very close to having an ability to carry out a very meaningful conversation with a computer system. Such systems will have an insight into vast libraries of scientific research and hopefully will draw new inferences from that material. The most spectacular applications might be related to medical sciences. For example, nanomedicine research projects are under way in all leading research centers around the world (Sahoo et al. 2007; Chih-Wei et al. 2011). Delivering medication to specific areas of human body, destroying tumor cells, or interfacing our bodies with non-biological computing platforms will be soon accomplished by nanotechnologies.

Those who are skeptical about the future capabilities of new technologies often point out that there are many unsolved research problems before general AI is created. Therefore, there are good reasons, at the moment, to dismiss the idea of a technological singularity. A dose of realism is always healthy. Nevertheless, our history is full of examples where skeptical realists were proven wrong. Pushing singularity into a distant future is another attitude. After all, putting new ideas into the realm of science fiction has been a common practice. It is certainly true, that in the field of AI there are many unsolved problems. The frame problem that deals with complicated knowledge representation issues (Dennet 2006), non-monotonic reasoning representing rules that might include exceptions (McCarthy 1984) are only a few examples of well-known algorithmic problems. The issue of unsolved research problems has been the subject of discussions in the AI circles for many decades. An interesting tactic proposed by some scientists suggests that focusing research on narrowly defined "toy" problems should be abandoned. Instead more emphasis ought to be placed on applied problems. Solving such problems as self-driving cars will provide a test bed for new generation of systems. They will be a stepping stone for development of other more intricate technologies.

Quantum computing is largely considered the key technology to keep the Moore's Law valid for decades in the future (Lundstrom 2003). The absence of fully functioning quantum microprocessors is a reason for skepticism regarding the future advancement of computational power. There are reports that laboratory experiments with quantum technology yielded promising results. Nevertheless, there are very few studies into how soon quantum logic gates can be manufactured. We have a much better understanding of how powerful our computers will become in this century based on theoretical analysis. The conservative estimates point out the end of Moore's Law. However, exascale computing[1] is a very real research topic in the academia and industry (Shalf et al. 2011) even if

Moore's Law is no longer in effect. Computerworld predicts a thousand-fold performance increase over the current petascale[2] supercomputers by 2018 (Computerworld 2010). This incredible performance increase seldom makes it to the headlines in the media. This level of computing power will be the enabling technology for modeling living processes and brain architectures (Kavanagh and Hall 2008). One can only hope that this research will not end up trivialized as much as the recent Mars missions or other great scientific and technological achievements (Muirhead 2004).

The computational power is a necessary condition for rapid progress toward general AI. The term general AI is increasingly used to describe the next generation of algorithms applied to support broadly defined tasks. The other key component is software. In order to harness the current parallel processing architectures, we need new approaches to system design and implementation. Researchers often doubt that software developers will be able to keep the pace with thousandfold increases in computing power availability. They make a valid point that system complexity will be approaching infinity, which is not a good environment for designing and writing code (Kunstler 2012). This point of view tends to overlook a huge body of software engineering research. Since the beginning of modern computing, founded on John von Neumann's work, one of the focal points of researchers is devising methods for handling inevitable increases of system complication (Selby 2007). These efforts focus on such areas as complexity of computational algorithms, development of new programming languages and design methods, software assessment tools, compiler design, and so on.

It is important to consider the opinions of unconvinced and skeptical scientists and engineers because we are talking about an event that is considered by many the most important one in human history: the creation of artificial intellects. Assuming that the current trends continue, in a few decades the available processing power will be millions of time faster than the processing power of our brains. Our understanding of complex algorithms and neuroscientific processes will be incomparably better. It is difficult to grasp the implications of that fact. Clearly, there are many paths leading to difficult and uncomfortable ethical, political, and economic questions.

The believers in technological singularity are by no means united in their vision. Many of them are convinced that reverse engineering of human brains is a good path. However, researchers are perfectly aware about the cellular level complexity and difficulties associated with modeling behaviors of billions of neurons and quadrillions of connections.

Also, extracting the "software of intelligence" from biological brains is a relatively new research field. Breakthrough discoveries might not materialize anytime soon.

3. New Technologies and Their Impact

The computational capability of the human brain or the nature of common sense reasoning has been always implicitly or explicitly present in AI research. Computational power of biological brains and common-sense reasoning remain to be difficult research questions. Scientists need answers to those and many more questions in order to establish the distance between the research objectives and the current state-of-the art. Above the engineering layer, there are equally challenging ethical and philosophical questions. How close the digital technology will be interfacing with humans and at what point digital technologies will start dominating human beings? What will be the impact of the subsequent cyborgization of the human kind (Zhao 2006)? The term cyborgization is slowly entering the scientific circles. It describes a merger of biological and non-biological technologies. Perhaps a human with mechanical hip replacement falls into the cyborg category. However, the future of mankind will be changed when we start interfacing brains with non-biological machinery.

Cyborgization of our species, that is, integration of non-biological technology and various body parts, became a reality decades ago. Artificial organ transplantology and joint replacement operations are almost routine procedures. We are entering a new era in which the integration of technological and biological systems starts affecting the spinal column and the brain. Scientists recently developed a technology that enabled a paraplegic to stand for several minutes and walk with assistance (Grose 2012). Building an interface between a computer system and a biological nervous system allows to take over some of the brain functions or to repair parts of the brain tissue. Such cyborgization technologies will be a tremendous benefit for people with disabilities. The applications are not limited only to helping patients with severe spinal cord injuries. There is research focusing on reading and supplying information directly to the brain.

Cyborgization, however, poses a number of ethical questions especially if we consider AI affecting higher cognitive functions. At what point, for example, a human being with his or her brain interfaced with a computer or AI becomes a different entity, that is, a cyborg? Can he or she be a human being and a cyborg at the same time? Who is responsible for cyborg's actions: the human or the maker of the AI system?

Estimating human brain computational capacity is an interesting problem. If we know the amount of computations that can be carried out in a biological neural network of that size—it will be an important point of reference. The parallel and analog architectures of human brains are fundamentally different from those of digital computers. Human brains are basically a highly interconnected neural network. Estimating the number of cells in the human brain is actually an interesting problem in itself. The literature on the subject of human brain typically puts the number of neurons at around hundred billion. This number might be merely a rough estimation and has been recently questioned in neuroscience literature (Azevedo 2009). There is no question, however, that we are looking at a network consisting of potentially over a hundred billion of nodes. Estimating the processing power of such a network can be done, at this point, only indirectly. A brilliant method has been proposed by Hans Moravec and discussed by Ray Kurzwail. Moravec postulated that we can analyze the retina, which has a very well-defined purpose: image processing. We have a rather good understanding of how much power a digital computer needs to process images. The number estimated by Moravec to be around 109 of instructions per second (1000 MIPS) describes the equivalent digital computer speed. Since the human brain is much larger than the retina, we can estimate that the entire brain is about 1014 instructions per second. This estimation is probably not perfect. But, even if it is off by a significant margin, it gives us an important point of reference for drawing parallels between biological and non-biological computations.

Very few specialists predict a hard limit for increases of computational power even though they point out limitations for our current technologies. Those limitations are mostly related to the power consumption and negative effects it has on the stability of super small circuitry. Quantum computing, while being a very promising technology, is still being researched in laboratories. Successful research into DNA-based computers, which could be made small enough to operate inside cells to control their activity, represent another path or research. At the moment, these technologies are predicted to revolutionize the pharmaceutical industry. This type of "classifier" biological microprocessor is used to detect cancer cells. Such a computer operates in a volume of a cubic micron and is not limited to biological tasks. The Science magazine recently featured a paper discussing general pattern recognition implemented in a "liquid logic" platform—a processor based on nucleic acid circuitry (Seelig et al. 2006). It is clear that general purpose computing with biological circuitry is seriously investigated.

There is a very valid point made by some analysts concerning algorithmic and programming foundations of AI. Increasing the processing power does not constitute a complete solution to the problem of constructing AI. Programmers must deal not only with the increasingly complex objectives like real-time image recognition or natural language processing. The hardware is increasingly complex as well. The Moore's Law is still in effect today because of the massive parallelization of supercomputers and cloud computing platforms (Rytter 1988). Parallelization is a way to handle computationally expensive problems without building super-fast and super-expensive microprocessors. Instead, a large number of off-the-shelf processing units can be networked. This way, computational tasks can be divided (or "forked") between multiple microprocessors. Each microprocessor works on a small chunk of the problem at hand. Once all jobs are completed, the final solution is generated. Not all problems can be decomposed into thousands of microprocessors. Scientists, however, are working on ways to improve models of parallel computation. It might be possible, for example, to alter any computing problem to make it well-suited for parallel computation. Parallel computation platforms based on OpenMP, MPI, Hadoop, and other technologies represent a major paradigm shift in the industry, research, and academia (Peng and Wang 2010).

The capability to execute algorithms in very fast computers is essential for almost all intelligent technologies to materialize. There are, however, other design and implementation issues if we move the discussion away from hardware and focus on high-level objectives. Common sense reasoning is one of the key components of the future human level artificial intellects. There are companies that are working on the solution since several decades. CYC Corporation has been founded to solve the commonsense reasoning problem by using rules of logic (Lenat 2001). They develop a huge knowledge base containing facts about our everyday life and environment. Common sense knowledge provides a relationship between such terms as "stone" and "heavy." There are millions of such rules that can be written. The process of creating such a knowledge base of common sense fact is a painstaking manual procedure involving teams of people entering these relationships in a consistent and organized way. After several decades of development, common sense reasoning is a tremendous accomplishment of the AI field. CYC knowledge bases are increasingly applied in military and civilian systems (Cohen et al. 1998). Systems with common sense reasoning capabilities will soon be driving our cars or will control our domestic robots (Urmson and Whittaker 2008).

It appears that the current line of brain scanning research generates very profound discoveries and might be a major step toward understanding processes behind human intelligence. The next step will be to replicate these processes in a far more powerful computational platform. There are many research initiatives focusing on understanding how biological brains process concepts expressed in natural language. Of particular interest is the brain's internal knowledge representation (Pereira et al. 2009). Studying visualizations of various brain activity patterns enables researchers to effectively read minds by associating visual activation patterns with words or more complex concepts. This fascinating research is addressing one of the fundamental problems in natural language processing: ambiguity of the natural language resulting mainly from synonymy, polysemy, and contextual relationships. Brain scans analyzed by scientists show that a specific concept activates different brain regions depending on the contextualized meaning.

The mind reading technology is a very frightening perspective because it has the potential to jeopardize the most fundamental privacy rights. We can imagine computer viruses infecting biological interfaces in order to perform scanning and recording of our thoughts. Such interfaces will be soon replacing keyboard and other input devices. We will be able to replace computer keyboards. Instead of typing, we will "think" the text or directives to be entered into the computer. Programmers will be able to write software by thinking about the code instead of laboriously entering thousands of lines of code through the keyboard. This technology will give artists a tremendous tool to visualize their creations by scanning them directly from their brains. It is only a question of time before systems recognizing visual patterns in human brains will become a hot research topic. On the other hand the mind reading technology offers important medical capabilities. People with spinal cord injuries will be able to walk by artificially reconnecting brain regions responsible for movement with external actuators controlling exoskeletons or other mechanical devices.

The increasing sophistication of the brain scanning technologies is a reality today. These tools will probably be crucial in the effort to reverse engineer the human brain. The brain scanning capability leads to a very difficult philosophical problem. It was always argued that only human brains have "real" feelings and a "real" understanding of the world that surrounds us. Digital computers, on the other hand, do not have any feelings—only binary impulses stored in various types of memory devices. Visualizations of human brain activation patterns accomplished by Tom

Mitchell's research group at Carnegie Mellon University constitute a result that is difficult to accept. There are many voices stating that we will never create systems that can truly understand the environment. Now, we might have a crucial piece of evidence that our "deep" understanding of concepts and feelings is nothing more than patterns and images of neural activations that can be displayed and analyzed with fairly simple scientific tools.

4. General Artificial Intelligence versus the "Real" Intelligence

One of the characteristics of the advantage enjoyed by AI researchers over scientists studying biological systems is based on the fact that their systems are not running on fixed hardware. Technologists can relatively easily increase the processing power and memory space of computers that execute the software. Unfortunately, human brain memory capacity and speed is fixed even though there might be some variability. Currently, the highly acclaimed AI system is Watson—a system created by a team of programmers and AI specialists at IBM. This system received fairly substantial but short-lived media coverage. Watson (Ferrucci 2010) successfully participated in the TV show Jeopardy where it was competing against human all-time champions. The format of the Jeopardy show is very difficult for a computer system. The breadth of knowledge required to successfully participate is enormous. The key to Watson's success was a deep understanding of natural language nuances. Jeopardy is known for challenging participants not only with broad topic selection. Often, questions presented to participants involve figures of speech, pun, or highly contextual statements derived from popular TV shows or movies.

The algorithms behind successful AI implementations are often based on statistical modeling of language. Watson is the most recent example. This is often a point used as a chief argument against further advancement of intelligent technologies. For some reason, mathematical or statistical models of knowledge or natural languages are considered to be an inferior toolset, one that is unable to capture the "real" knowledge. In such discussions people tend to forget that human brain activation patterns can be accurately modeled by statistical functions. So, in fact, we can compare a mathematical model of a concept retrieved from a vast database to a model of the same concept derived from human brain patterns. Can we characterize one of them as fundamentally better? It is difficult to refuse the statement that intelligence is an emergent property of brain activities. But, it is simply inconsistent to dispute the validity of mathematically modeled patterns. The argument that only human brains can build associations between large numbers of loosely relevant facts has been proven wrong. Watson and many other systems utilizing similar algorithms represent a great step toward general AI based on reasoning algorithms and large knowledge bases. Is Watson a model for future progress in AI? At the moment the answer is yes. Research in machine learning, reasoning, logic, and other fields of AI is truly impressive. Research in such a mature field as AI is difficult to follow without a lot of dedication. However, such scientific conferences as The International Joint Conference on Artificial Intelligence provides a window into the inspiring future of intelligent systems (IJCAI url 2012).

Artificial intelligence research has been rooted in computer science and mathematics since the 1950s when Marvin Minsky and John McCarthy coined the term "artificial intelligence." Recently, however brain research conducted in fields of neuroscience appears to be increasingly important. In 2008, there were about 50000 scientific papers published in the field of neuroscience. The number of brain research publications is doubling every year (Baker 2011). Every aspect of research about modeling and simulating human brain is accelerating. One of the directions is to interface human brains with computers. There are numerous biological challenges that have to be solved. Neurons typically isolate foreign objects but surgically installed neural implants have been successfully functioning. Transferring our "brain software" to computer systems might emerge as a primary direction of neuroscience research (Swan and Howard 2012). Whole brain emulation (WBE) is likely to be the research platform influencing architectures of computer systems (Deca 2012). Computing architectures established on the basis of von Neumann's designs are modified by brain researchers (von Neumann 1986). It is important to note that biological systems are a great source of knowledge for researchers studying principles of computation. Ironically, biological brains are suboptimal in terms of speed. The speed of communication between neurons is very slow compared to semiconductor-based systems. This situation is perceived as an opportunity for achieving higher levels of intelligence because speed of computations is considered a less difficult hurdle to overcome.

Artificial neural networks, well-established tools in the field of machine learning, are an important technology for the future of intelligent computing. Systems based on artificial neural networks are perfect devices for developing classification models of highly non-linear patterns. Such non-linear problems are face recognition or character recognition tasks. Document classification algorithms based on artificial neural networks are capable of capturing intricate semantic relationships in very large databases and provide question answering capabilities. Such remarkable capabilities are accomplished with relatively simple means. Architectures or artificial neural networks are incomparably simpler than those of biological brains. In the context of our constantly improving understanding of biological neural systems, artificial neural networks will probably be an algorithmic foundation for high-speed brain emulation systems.

We could discuss many more examples describing sophisticated AI technologies. In almost all cases, as soon as an AI technology is successful, it becomes a routine academic discipline. This was the case with robotics, data mining, or machine translation. The new breakthrough technologies like self-driving cars controlled by complex artificial intellects or household robots are already trivialized. However, these technologies are firmly entering the mass production phase and will be available on the market within a decade in many consumer products (Urmson and Whittaker 2008; Markoff 2010).

Artificial intellects will sooner or later exceed human intelligence. Machines can quickly share huge amounts of information which is increasingly available in digitized format. Memory capacity and computational power are rapidly increasing in the cloud architectures currently available. Intelligent system will be able to access their own blueprints and implement improvements in the design of hardware and software. These tasks can be accomplished today although the projects are typically implemented in research centers. One of the main important areas of focus of machine learning community is to make the machines learn in unsupervised or semi-supervised fashion, without interruptions and over long periods of time. If these evolutionary processes start outside tightly controlled laboratory settings, AI will become a runaway phenomenon of rapidly improving super-intelligence.

5. Political and Social Impact

Technological rivalry between political superpowers is about to enter a new phase defined by the emerging intelligent technologies, neuroscientific knowledge, nanotechnologies, and robotics. Philosophical and theological research into AI and transhumanism might lack the prestigious funding from governmental agencies but it touches very profound problems affecting the global economy and military ambitions of many governments. One of the biggest questions implicitly embedded in the big picture of the emerging technologies is: do we want to fully exploit the emerging technologies? Similar question was asked during the course of the Manhattan project by physicists who realized the destructive power of nuclear energy (Smyth 1945). The path toward technological singularity is equally dangerous. This path has been played down by the sci-fi movies and TV shows or placed in the very distant future. Nevertheless, the economic and military pressures paired with the scientific momentum are very real. They lead to billions of dollars invested into military robotics and AI-based weapon systems. Soon we will be deploying autonomous military systems. There will be inevitable accidents involving military and civilian AIs behaving in an unexpected but clever way. Manufacturers around the world devise increasingly complex robotics and automations systems removing humans from factories which leads to increasing unemployment among factory workers. The future involving totally unmanned production facilities performing complex manufacturing tasks and maintaining the facilities by other robots is similarly unsettling for many people. Unmanned drones are about to start monitoring the streets from the air invading our privacy. The social and political movement to limit the research and widespread applications of intelligent technologies will become an important issue in the upcoming election cycles.

The less publicized objective for new technology development is to ensure that the intelligent systems do not malfunction, that is, stop functioning if something goes wrong. One of the focal points in military research is development of "self-healing" software (Khan 2011; Blair 2002). It will be interesting to see political and public reaction after the inevitable cases of robotics systems surprising the public by exhibiting behaviors reflecting their own agenda. By design, these drones will behave in a way that shows resilience to destructive forces and at the same time capabilities of developing clever tactics to avoid deactivation.

Household robotics, however, is just around the corner (Denning 2010). Humanoid robotic systems for assistive elderly care will probably enter hospitals and retirement homes first (Broekens et al. 2009). After a while, each software upgrade or hardware replacement resulting in a better vision system or faster computer control system will make the robots appear smarter and better to a point of anxiety for many people.

Hugo de Garis discusses important consequences of this anxiety in the political and social context (de Garis 2005). De Garis predicts several scenarios that are likely to play out in political and social arenas. The methodology of his discussion is not particularly surprising. There are many historical examples pointing to similar circumstances at times of revolutionary transformations. At the root of the discussion are three fundamental social and political attitudes toward AI systems.

The first attitude will be characterized by strong opposition toward artificial intellects. Some people will perceive the powerful artificial intellects as a thread that will disenfranchise millions of workers or control a super weapon with its own agenda. Followers of this attitude will advocate that intelligent technologies be controlled or delegalized in a way similar to current laws regulating weapons of mass destruction. Probably, these efforts will not be entirely successful. If this point of view prevails, artificial intellects will not be accessible in household consumer electronics but military systems will surely continue to evolve into increasingly capable entities. AI has been noticed by policy experts. Preparations for legislative initiatives addressing effects of new technologies on the battlefield and within our society are already under way (McDaniel 2008). Similarly in Europe, The European Robotics Research Network funded research on roboethics from the point of view of the designer and the user (Veruggio 2006).

The second attitude will be represented by people who embrace the new technologies and will see it as an opportunity to overcome our biological limitations. The posthumanist ideas are very tempting. Artificial intellects working side-by-side with humans or as systems interfaced with human brains will help solving difficult scientific problems (Antipov and Koldomasov 2010). The transhumanist groups will eagerly embrace the cyborgization of our society anticipating breakthrough scientific discoveries made be super-intelligent systems. Discoveries in genetics and energy sources have the potential to transform our society in positive ways by eliminating diseases, understanding and solving the energy crisis.

Finally, the third attitude intersects the transhumanist ideas with religious movements. Super-intelligent artificial intellects will be perceived very God-like in the eyes of many. This point of view is outlined by Jaron Lanier's cybernetic totalism theory which states that the universe is basically an information pattern (Lanier 2011; Jackelén 2002). Quantitative improvements in computing technology will lead to equal qualitative improvements which have the potential to create God-like entities. After singularity occurs AI promises super-intelligent machines and immortal minds.

6. Summary

Figuring out how our complex and chaotic biological brains work will take many decades. We will certainly reach this goal but it will be a process involving many smaller discoveries. Software engineers, mechanical engineers, neuro-scientist, nano-technology experts, and many others are collaborating on many exciting projects in robotics and biological sciences. Perhaps our mathematical tools will turn out to be insufficient and evolutionary engineering tools based on "mutate-test-select" approach will be the key. However, it is only a question of time before humanity will start embracing the enormous potential of the new technologies.

Notes

1. Exascale computing describes computational power a 1000 times faster over the current strongest supercomputers.
2. Petascale is a term describing computational power of today's fastest computers, that is, one quadrillion floating point operations per second.

References

Antipov, M. A., and A. S. Koldomasov. "Cyborgization of Mankind as a Display of Transhumanism." *Sociosfera* 4 (2010).

Arnol'd, V. I. *Singularity theory*, vol. 53. Cambridge, MA: Cambridge University Press, 1981.

Azevedo, F., et al. "Equal Numbers of Neuronal and Nonneuronal Cells Make the Human Brain an Isometrically Scaled-Up Primate Brain." *The Journal of Comparative Neurology* 513, no. 5 (2009): 532–41.

Baker, S. *Final Jeopardy: Man vs. Machine and the Quest to Know Everything*. Boston: Houghton Mifflin Harcourt, 2011, 166.

Blair, G. S., et al. "Reflection, Self-Awareness and Self-Healing in OpenORB." In *Proceedings of the first workshop on Self-healing systems*. ACM, 2002.

Bostrom, N. "A History of Transhumanist Thought." *Journal of Evolution and Technology* 14, no. 1 (2005): 1–25.

Broekens, J., M. Joost, M. Heerink, and H. Rosendal. "Assistive Social Robots in Elderly Care: A Review." *Gerontechnology* 8, no. 2 (2009): 94–103.

Bruce, J. W., and P. J. Giblin. *Curves and Singularities: A Geometrical Introduction to Singularity Theory*. Cambridge, MA: Cambridge University Press, 1992.

Brynjolfsson, E., and A. Saunders. *Wired for Innovation: How Information Technology is Reshaping the Economy*. London: MIT Press, 2009.

Chacko, D. M., et al. "Transplantation of Ocular Stem Cells: The Role of Injury in Incorporation and Differentiation of Grafted Cells in the Retina." *Vision Research* 43, no. 8 (2003): 937–46.

Chih-Wei, C., K. H. Chou Hou, L. Shieh, H. Sheng-Hsin, and J. C. Chiou. "Wireless Powering Electronics and Spiral Coils for Implant Microsystem toward Nanomedicine Diagnosis and Therapy in Free-Behavior Animal." *Nanoelectronics Conference (INEC), 2011 IEEE 4th International*, 2011, 1–2.

Cohen, P. R., et al. "The DARPA High-Performance Knowledge Bases Project." *AI Magazine* 19, no. 4 (1998): 25.

Computerworld. "Scientists IT Community Await Exascale Computers, 2012." http://www.computerworld.com/s/article/345800/Scientists_IT_Community_Await_Exascale_Computers, viewed on December 15, 2012

De Garis, H. *The Artilect War: Cosmists vs. Terrans: A Bitter Controversy Concerning Whether Humanity Should Build Godlike Massively Intelligent Machines*. California: ETC Publications, 2005.

Deca, D. "Available Tools for Whole Brain Emulation." *International Journal of Machine Consciousness* 4, no. 1 (2012): 67–86.

Dennett, D. C. *Brainstorms: Philosophical Essays on Mind and Psychology*. Cambridge, MA: MIT Press, 1981.
Denning, T., et al. "A Spotlight on Security and Privacy Risks with Future Household Robots: Attacks and Lesson." In *Proceedings of the 11th International Conference on Ubiquitous Computing*, ACM, 2010.
Esnaola, U., T. Smithers, and I. J. Agirre. "A society of agents for service robots." Industrial Electronics (ISIE), 2010 IEEE International Symposium, 2010, 3515–20.
Ferrucci, D., et al. "Building Watson: An Overview of the DeepQA Project." *AI Magazine* 31, no 3 (2010): 59–79.
Ferrucci, D. "Build Watson: An Overview of DeepQA for the Jeopardy! Challenge." In *Proceedings of the 19th International Conference on Parallel Architectures and Compilation Techniques*, ACM, 2010.
Gelsinger, P. "Moore's Law—The Genius Lives On." *Solid-State Circuits Newsletter, IEEE* 11, no. 5 (2006): 18–20.
Grose, T. K. "Up close: Shock and Awe." *Prism-Washington* 22, no. 1 (2012): 24.
Hassler, C., T. Boretius, and T. Stieglitz. "Polymers for Neural Implants." *Journal of Polymer Science Part B: Polymer Physics* 49, no. 1 (2011): 18–33.
Hobsbawm, E. J. *Industry and Empire: The Birth of the Industrial Revolution.* New Press, 1999.
IJCAI. http://ijcai13.org/, accessed on November, 2012.
Jackelén, A. "The Image of God as Techno Sapiens." *Zygon* 37, no. 2 (2002): 289–302.
Kavanagh, J., and W. Hall. "Grand Challenges in Computing Research." 2008 in http://www.ukcrc.org.uk/grand-challenge/, accessed on November 2, 2012.
Khan, A. "Self Healing Software." U.S. Patent No. 8,082,471, 2011.
Koh, H., and C. L. Magee. "A Functional Approach for Studying Technological Progress: Extension to Energy Technology." *Technological Forecasting and Social Change* 75, no. 6 (2008): 735–58.
Kunstler, J. H. "Too Much Magic: Wishful Thinking, Technology, and the Fate of the Nation." *Atlantic Monthly Press*, 2012, 69.
Kurzweil, R. "The Singularity is Near: When Humans Transcend Biology." *Viking Adult*, 2005, 67.
Lanier, J. *You are not a Gadget: A Manifesto*. New York: Vintage, 2011.
Lenat, D. B. "From 2001 to 2001: Common Sense and the Mind of HAL." 2001. www.cyc.com/cyc/technology/hals-legacy.rtf, accessed on December 15, 2012.
Lloyd, S. "Computational Capacity of the Universe." *Physical Review Letters* 88, no. 23 (2002): 237901.
Lundstrom, M. "Moore's law forever?" *Science* 299, no. 5604 (2003): 210–11.
Markoff, J. "Google Cars Drive Themselves, in Traffic." The New York Times 10, 2010, A1.
McCarthy, J. "Some Expert Systems Need Common Sense." *Annals of the New York Academy of Sciences* 426, no. 1 (1984): 129–37.
McDaniel, E. A. *Robot Wars: Legal and Ethical Dilemmas of Using Unmanned Robotics Systems in 21st Century Warfare and Beyond.* Fort Leavenworth, KS: Army Command and General Staff Coll., 2008
Merlet, J. P. "A Formal-Numerical Approach for Robust In-Workspace Singularity Detection", *Robotics, IEEE Transactions* 23, no. 3 (2007): 393–402.
Minsky, M. *Society of Mind.* New York: Simon & Schuster Paperbacks, 1988.
Moore, G. E. "Cramming More Components onto Integrated Circuits." reprinted from *Electronics* 38, no. 8 (1965): 114 ff. *Solid-State Circuits Newsletter, IEEE* 11, no. 5 (2006): 33–35.
Muirhead, B., and K. Brian. "Mars Rovers, Past and Future." *Aerospace Conference Proceeding, IEEE* 1 (2004).

Nelson, R. R., and E. S. Phelps. "Investment in Humans, Technological Diffusion, and Economic Growth." *The American Economic Review* (1966): 69–75.

Peng, Y., and F. Wang. "Cloud Computing Model Based on MPI and OpenMP." In *Proceedings of Computer Engineering and Technology (ICCET), Vol. 7, IEEE,* 2010.

Pereira, F., T. Mitchell, and M. Botvinick. "Machine Learning Classifiers and fMRI: a Tutorial Overview." *Neuroimage* 45, no. 1 (2009): 199.

Rytter, W. "On Efficient Parallel Computations for Some Dynamic Programming Problems." *Theoretical Computer Science* 59, no. 3 (1988): 297–307.

Sahoo, S. K., S. Parveen, and J. J. Panda. "The Present and Future of Nanotechnology in Human Health Care." *Nanomedicine: Nanotechnology, Biology and Medicine* 3, no. 1 (2007): 20–31.

Selby, R. *Software Engineering: Barry W. Boehm's Lifetime Contributions to Software Development, Management, and Research.* Wiley-IEEE Press, 2007, 797–806.

Shalf, J., S. Dosanjh, and J. Morrison. "Exascale Computing Technology Challenges." *High Performance Computing for Computational Science-VECPAR 2010*, 2011, 1–25.

Smyth, H. DeWolf. *Atomic Energy for Military Purposes*, vol. 4, Princeton: Princeton University Press, 1945.

Stackhouse, B., et al. "A 65 nm 2-Billion Transistor Quad-Core Itanium Processor." *Solid-State Circuits, IEEE Journal* 44, no. 1 (2009): 18–31.

Seelig, G., et al. "Enzyme-Free Nucleic Acid Logic Circuits." *Science* 314, no. 5805 (2006): 1585–88.

Swan, L., and J. Howard. "Digital Immorality: Self or 0010110?" *International Journal of Machine Consciousness* 4, no. 1 (2012): 245–56.

Urmson, C., and W. Whittaker. "Self-driving Cars and the Urban Challenge." *Intelligent Systems, IEEE* 23, no. 2 (2008): 66–68.

Veruggio, G. "The Euron Roboethics Roadmap." *Humanoid Robots, 6th IEEE-RAS International Conference, IEEE,* 2006, 612–17.

von Neumann, J. *Papers of John von Neumann on Computers and Computer Theory.* Cambridge: MIT Press, 1986.

Zhao, S. "Humanoid Social Robots as a Medium of Communication." *New Media & Society* 8, no. 3 (2006): 401–19.

Part Three

Good Robots—Philosophical Perspective

Robot Desires

Timo Airaksinen
Helsinki University, Finland

1. Abstract

If we cannot design robots that are like humans, we will design humans that are like robots. Moreover, robots need humans and humans need robots. This looks more symmetrical than it actually is. How symmetric are these relations? What we actually do is design a culture in the technological frame, and within such a frame, we live the lives of cyborgs. The word of choice here is "design," not "creation," "production," or "manufacturing," because we may never realize our designs. I will explain what this means in reference to novel desires or machine dreams. As examples, we can use airports, the existing military drones, and fictional domestic nanny robots. The methodology of this paper is praxeological even when it combines fact and fiction.

2. Introduction: Robot Wars

It is the year 2465 in Nova Philadelphia, twenty years after a war when robots rebelled against their human masters and destroyed a large part of them. The humans fired their nuclear weapons disabling most of the robots, but at the same time, they hopelessly crippled the planet Earth. Then suddenly and mysteriously the robots shut themselves off. The humans destroyed the robots and declared a moratorium on the development, use, and application of all electronic devices. However, one robot still exists in working condition, the C-12, a household humanoid nanny robot. What should they do with it/her? In the end, even this last robot is disabled but not dismantled, and Penelope, a young, crazy techno-buff, will restart it/her. What are the consequences of her recklessly illegal decision? Such a technological melodrama does not describe our real world, but it is fascinating to try to see where we are going.[1]

Claude Shannon once devised a robot called the Ultimate Machine which, incidentally, greatly impressed Arthur C. Clarke when he saw it in action. The initial idea came from Marvin Minsky in 1952. The robot is just a box that has a switch on top of it. When you throw the switch, the box makes a buzzing sound. After a while a lid on the box opens and a hand comes out. This hand reaches for the switch and switches off itself/himself. Then the hand retreats and the lid closes. There is silence. The robot has committed suicide and fulfilled its unconscious thanatotic desire, as Lydia Liu freely interprets the scenario.[2] Does this make sense philosophically? Can robots really have desires that motivate them consciously and unconsciously? I first present a skeptical argument and then move on to argue that desires have their place in the correctly understood technological world of robots.

3. Our Needs, Their Needs

The question is, do robots as particular machines have needs and desires? What I mean by a *particular* robot is a non-human machine that possesses a high degree of artificial intelligence, is capable of self-movement, has sensory capabilities, and preferably has hand-like extensions for working.[3] If such a thing also has desires, we may consider it as approaching the status of an individual. I will argue that robots may not have desires or full-blown intentions for that matter. I will then explain in which sense desires can penetrate the robotic world. However, we must address the problem of needs first. It seems that particular robots have needs just as humans have needs. Needs are essential life- and welfare-related demands whose satisfaction allows us to survive in a hostile world.[4] We humans need water and food, shelter, and clothing. We need company and some basic skills. We need many things if we are going to survive. Here, we talk about basic needs, of course, not needs like instrumental needs, such as our need of a bicycle in order to have fun. Instrumental needs are grounded on desires; hence, they are not basic. What I mean is that you do not need a bicycle if you do not want to have fun. Basic needs do not refer to desires in the same way. When you need water you think of drinking, it's not your desire to pleasantly hydrate yourself. Basic needs do not mention desires.

Robots have their needs, too. They need shelter, oil, energy, maintenance, and spare parts. They do not need loving care, advice, or polishing, as these are not fundamental requirements of their continuous functioning in an effective and efficient manner. You may like to caress them, but this is another thing. If they do not function, we simply scrap them, or perhaps some technology buffs display them in their collections, as Penelope did. However, as long as a robot works and functions, it has its needs that must be recognized and respected. A modern vacuum cleaning robot finds its way back to the energy source in the room, and then after recharging, it will continue its toil. A modern car tells its owner when it needs its periodic maintenance by giving an alert. The human agent is supposed to react to this alert by answering it. Then the robot is happy and confident again, as one may say.

Unlike desires, needs as facts are extensional entities. They are not intentional states of mind like desires. Some people think that human needs are qualitatively different from machine needs, but this may well be because of confusion between needs and desires. If we keep them apart, humans are biomachines that are in need of energy and maintenance just like robots, as machines, are. When a working robot shows signs of malfunction, you fix it because it needs repairs. You do this without raising any issues concerning its conditional needs. You may not think if I want the machine to keep functioning, perhaps I should fix the problem. You simply fix it. In the same way, you do not ask if I like to keep my (human) workers healthy, I should fix, say, the heating system and air filters. The robots and the (human) workers have their basic needs and that is that. Of course, the ethical consequences of neglecting the human needs are more serious than neglecting the machine needs, but this is another matter. I agree, though, that we should not talk about the happiness or welfare of robots. That would be an inappropriate ethical metaphor. Yet, both humans and robots need effective repairs to make them work efficiently.

4. Human Desires

Notice how complicated mental state desires are. I present a rudimentary model of human desire and ask then whether it is possible for a robot, a highly advanced and perhaps fictional robot for that matter, to possess these mental states. A human state of mind, a full-blown *desire*, contains the following elements: imagination, thought, wish, hope, want, intention, choice, experience, welfare or happiness related consideration, and value.[5] For instance, I desire to learn Latin. This begins as a thought and an idea in my imagination, it is then a wish in my mind when I consider it further, I recognize that I want to do it, and then I intentionally commit myself to the project. Next, I choose my teacher and ask her for advice. I get a learning experience which makes me happy and is valuable to me. Such a sequence requires certain beliefs as epistemic states and a suitable social environment, but I skip these external factors in this brief paper. No robot can possess more than a thought, if we are willing to accept a lenient interpretation of the relevant facts. All other ascriptions of the relevant mental sub-states are just bad fiction or clumsy metaphors.

The situation gets worse when we consider the elements of desire as *mimetic* states, metonyms, and narratives. Human desires are mimetic in the sense that a person's desires reflect and copy other person's given desires.[6] To put it bluntly, we all desire the same thing. Robotic designs may imitate it; all robots want the same thing because they copy their actions from one paradigmatic model. However, it gets worse. We use terms *metonymously* when a part represents the whole, as Hollywood represents the American film industry or "Give me a hand" represents whatever helps me in trouble. Why are desires metonyms? The problem with desire is that we often repress and deny our thoughts, as they are impossible, impermissible, shameful, unethical, and bad. We have difficulties with many of our images and we try to avoid them. In psychoanalysis, the main source of desire is the oedipal function which, of course, we must repress. You cannot deny the relevant desires, and so you transform them into something more acceptable and innocent. Therefore, the lustful male desire of one's mother experiences a metonymic transformation into something else, and the same happens when you desire to kill your father.[7] You love your books, art, and your work. You hate violence and discord. These new, acceptable desires can then be subjected to the mimetic process that copies them throughout the social world making them generally comprehensible and acceptable. Not all desires manage as well. Love may transform itself in unacceptable ways that allow only a partial mimesis. My love, which is originally oedipal, finds its metonymic expression in women's shoes which, as erotic objects, are not fit for the mimetic process. They remain as the objects of my personal, idiosyncratic, and perverted desire. It may not be too interesting to present such a psychological theory and then say that robots cannot be designed to model the same theory. The narrative involved is too complex for that, and thus, the critical argument is self-evident. However, this is not my main point.

5. Transhumanism, Cyborgs, and Robots

Transhumanism says that human beings can and will be modified by means of biological and technological means until we are perfect.[8] What we mean by *perfection* in this context is problematic, but that may not worry us now. Let us say that we become better and better without limit until we are moral, intelligent, wise, healthy, and nearly immortal. To some people this is an attractive dream that is gradually becoming a desire which entails certain choices in life and science and its accompanying values. We want to be happy, and this is the most efficient way of making us all happy in the world that is devoid of any hint of an otherworldly paradise and the mercy of benevolent gods. The transhumanist project is our only hope in the future. I have no opinion about this ideology. It is true that education, medicine, and technology can make us better; but are we really going to aim at perfection? Is it in our reach? These are demanding questions that do not relate directly to the idea of making human beings better. The reason is that perfection may imply the change of human nature into something new, for instance, a *cyborg* or a machine-enhanced pseudo-human being. We design a robot by starting with human beings. We do not make us better, we change ourselves.

Cyborg-related thought and ideology has its two branches: wide and narrow or particular and universal. As Donna Haraway has argued, we already live in a cybernetic cyborg world that is dominated by technology, technological thinking and minds, and human beings who are essentially connected to that artificial, engineered, and designed world.[9] It is a cybernetic world and at the same time reality without cyborgs. Let me explain. Cyborgs are normally understood as some kind of semi-robots that, unlike robots, are partly human and remain as such regardless of what technological modification they go through. If we say that these cyborgs are particular human-machine combinations we discuss the *narrow* perspective. Here, just like when we discuss robots, we can argue that however much we improve humans by means of technology, it does not matter. The resulting designs are still human beings with their typical beliefs, desires, and narratives.

What we cannot design and create by artificial means is our psyche, and this applies as well to cyborgs as robots. We may start from two opposite directions and then reach the same impasse. When we design robots, we start from simple machines and make them more complicated by giving them artificial intelligence, motion, perception, etc. Yet, we cannot transform them into something that could compete with humans, such as desires. When we design cyborgs, we start with a human being. We add technology, artificial intelligence, sensory aids, and so forth, without any preconceivable limits. Yet, we will not succeed in making this cyborg something that is more than a human being. The reason is that we cannot design a new psyche for it/her, and as long as we are keeping to the old one, however much we have enhanced its capabilities, we still have the same entity in our hands, a human being. In sum, you may start from a machine but you cannot design a being that is equivalent to a human being; you may start from a human being and make him a cyborg, but you can never eliminate the human characteristic of this new entity. Human psyche remains a fixed point in both of these lines of design.

Nevertheless, a way out of this dead end exists if you want to pursue the transhumanist line of thought. As I said above, we can adopt either a wide or a narrow concept of cyborgs. Earlier, I discussed the narrow approach that sees that cyborgs are particular entities in space and time. The wide approach understands our whole culture, our life-world, and scientific technology to be a cybernetic world of cyborgs. In other words, we humans are subject to transformation into cyborgs simply because we live in an artificially enhanced cyborg reality.[10] We are humans when we live in our typical social world; for instance, we imitate each other's desires, and we share our dream- and desire-related narratives. In the cybernetic world, all is different. We do not imitate each other's desires because we are programmed to desire. We do not dream the narratives that ground our desires. We watch them while we remain passive. We have only rudimentary desires which we confuse with needs. Our needs are satisfied, yet we do not know why and how, but we do not care anymore in the cybernetic world. It is a virtual world anyway for the most of us, and this world is a happy one. For example, many people already live on the web, by the web, and for the web. Hence, they are cyborgs in the wide sense of the term.

6. Heidegger: Frame and Resource

Martin Heidegger draws a distinction between technology as a resource and a frame in his philosophy of technology.[11] I use his illustration. Think of an airplane parked on a runway. This machine, an airplane, is designed to do certain work efficiently and effectively. It really flies and carries passengers, it minimizes the effort of traveling, and it is an economical and safe machine. It increases our happiness and contributes positively to our well-being. The airplane is a praxeological *resource* for us. It sits and waits until we desire to use it. It has no soul of its own, no desires, and no purpose until we assign value to it. It is just a simple resource, simple in spite of its wonderful technological complexity. Notice that this does not entail its moral complexity or the complexity of our desires in relation to it. Its essence is simple although its appearance is truly complicated. We need not admire its essence, travel with minimal effort, although we should admire its design and all the work and ingenuity that went into its design as a technological miracle. In essence, it is just an efficient and effective resource.

Such a resource has its dark side which may make us skeptical about its claims to effectiveness or even ethics. The problem is that technology does not leave the world as it has been but transforms it into something else, that is, resource. The scenery where the airplane is located becomes a resource called an airfield. We level and pave the land and then paint and illuminate it. An airfield is where the planes go and where they park. The sky above becomes a resource too, namely, airspace where the planes find their planned routes and which we reserve for them and where any birds become a hazard. We should keep in mind that modern technology does not mean particular machines, but it does mean their constellations and multiple interconnected systems of such constellations. Heidegger himself wrote about technology so early that he did not see clearly this aspect of modernity. Technology is a system of systems; and thus, it functions as a totality that replaces what has been.

The second aspect of technology is a *frame*. Heidegger tries to explain what this means, but in the end his reader must explain it to himself. This is how I see it. Technology is a frame in the sense that its system envelopes and contains us in the modern world now and in the future. It created a horizon against which all our desires, choices, plans, and hopes make sense, if they make sense at all. What I say is that our desires become technological; they become machine dreams which we entertain in this given frame whose name is technology. The frame starts looking like a necessity, a compulsory thing that we must accept. Yet, it is not enough to accept it; we must desire it. It is all good, valuable, and desirable as it enhances our well-being and contributes to our happiness. Whatever its problems are we can handle them by technological means. We have resources to do that, too. In this way, the problems of technology do not refer outside of its own frame. On the contrary, it is all immanent. The system and the frame are all encompassing formulations of a new cybernetic techno-universe.

7. Robot Desires

Robots are unable to desire, or they would no longer be robots. Cyborgs will never lose their desires; otherwise, they would not be cyborgs or technologically enhanced human beings. A similar point applies to life. When something is alive, it finally dies; so life leads to death, and so new life emerges. When Frankenstein creates his monster, what happens? As the story unfolds, the genius of Ingolstadt makes the dead body parts that form the monster and become alive. This is not accurate. What happens, in fiction at least, is that the dead body wakes up to become *undead.*[12] Undead is not the same as being alive. Undead is what emerges when a dead thing ceases to be dead. This is, of course, a fictional case. The problem that we need to solve here is this: when a dead thing wakes up, it is necessarily and automatically something *uncanny* (*unheimlich* for Freud) which entails fear and panic.[13] It is not only outlandish, it is uncanny in a much stronger sense. Think of Shannon's Ultimate Machine, the hand that switches itself/himself off and returns back. It is indeed uncanny as it seems to be undead. Think of the Golem, automated artificial limbs, moving puppets, talking dogs, zombies, and all those artificial designs that used to amuse and horrify audiences all over the world when they were not yet as sophisticated and bored as we are today.[14] Nevertheless, the ideas of living and undead tend to fascinate and horrify us to this day. Try to imagine a cyborg modification of a person who has a full mechanical mouth when false teeth, alone, are strange enough.

This leads to the following result. We live in a technologically defined and designed cybernetic world where we are cyborgs, in the wide sense, because of our techno-environment as a frame. This frame contains robots in addition to us cyborgs. The robots may be primitive or as sophisticated as you dare to imagine. The only requirement is that you may not imagine a C-12—a full-fledged android or humanoid robot. You may think of domestic nanny robots feeding babies or drones hovering over the tribal lands of Pakistan ready to kill or the intensive care unit of a hospital full of fancy robotics in the service of the sick and dying. None of those robots care, nor do they want anything. They only do their own work, given that their needs are satisfied in time. Yet, they create a technological frame in which our cyborg desires are situated. Such desires are complex, detailed, far-reaching, creative, motivating, and truly human. What will happen to them when we discourse with robots in a world that technology frames to the extent that we can call it a cybernetic world? Our desires will be uncanny.

Robots and cyborgs can never unify their diverse essences; they never meet just like the living and the undead will never meet. This sounds like an incongruity in need of its solution, an uncanny valley, an essential fracture over which we need to design a bridge, or it will haunt us forever. This is more than a matter of linguistic ambiguity. We do not know what is human and what is machine, we do not know what is alive and what is undead, and this is unlike it should be. We seem to have no solution to the problem of what is uncanny. Sigmund Freud argues that our castration complex creates the mental problem here. In the case of the Ultimate Machine, the hand that has no body lives its own separate life and so it is undead. The hand cannot be alive, as we realize, but it is not dead either. Yet it/him commits suicide as if it/he could decide by itself/himself, as if it/he had desires. This is uncanny as an instance of castration complex, and the separated hand is an instance of this. This must be nonsense, I am sure, although it may sound like good science fiction. What else could it be? The hand is a metonym of a comprehensive undead robotic body; its part that represents the whole. As such, the desire to commit suicide becomes comprehensible. It is not just the hand but also the whole creature who wants its own death or perhaps nirvana according to the Buddhist lore. Thus, the hand goes back to its box, or coffin, to rest in peace unperturbed by the external world. No ambiguity bothers us here. Yet, the uncanny aspect has not vanished. The image of the hand still troubles me as an instance of robot desire. Here, the material instance of a metonym is indeed uncanny as it seems to express a given desire which is too human to be true.

Robot desires are (almost) comprehensible in the robot world in their own technological frame, not as particulars, not as tokens but types, not really but as narrative elements. What happens to the human desires in this divided world? Whose desires will win? Those of the techno-robots or those of the human-cyborgs? As I said, only humans can have human desires, which is a mere platitude. But no one knows what desires we will have in the brave new world. Lydia Liu mentions oversimplification, automatism, and predictability as their constraints;[15] we can add repetition, alienation, boredom, control, surveillance, certainty, and lack of perspective in order to characterize this world where we cyborgs live a robot's life together with the robots. In other words, robot desires are fictional, but they may still serve as a model for us human cyborgs if we are unlucky enough. It is uncanny that we living human beings should imitate the undead robots in the technological frame.

Here is a story. The whole world, for instance, is an enormous high-tech international airport like Copenhagen or Airoport del Prat in Barcelona. It is stylish, clean, efficient, beautifully designed, shining, and so forth. It is located in the middle of nowhere outside the big city. It has a definite location but you have no idea where it is. When you look out of its glass walls you see airplanes, workers, trucks, and runways that extend all the way to the horizon. The terminal is isolated in an unspecified location as if it were nowhere but yet, at the center of the world.[16] It is luxurious to the desiring eye. Its shops and restaurants are endless just like its runways are. They are full of expensive, fine goods ready to flirt with you if you so desire. The food is spectacular, too. There is caviar, lobster, kosher salami, fresh fruits, fine drinks, the best cognacs and malt whiskies, and luxury wines in locked cabinets; also included is a walk-in humidor and designer clothes. It has everything you may ever desire. This is a paradise within a machine frame.[17] Why did the Shannon's hand go back to the box when it could have grabbed all this?

All this is uncanny, ambiguous, and castrated. I mean these desires and all the prospects of life. The airport is a machine and a cybernetic world fit only for robots. Alas, I am not a robot; at most, I am a cyborg, as my desires are intact. Yet, I am supposed to live a life that is standardized, regulated, regimented, repetitive, and normalized. This is life fit for the undead, an uncanny experience. All choices and desires have been previously defined, calculated, estimated, and predicted. It is statistical. All this starts looking like a robot world where even a cyborg must feel castrated and increasingly desperate.[18] Is there any place here where a cyborg can enjoy itself/herself as a unique individual human being? The answer is in the negative because CCTV surveys its/her behavior, and the robots who watch you do not understand aberrations or unpredictability as little as the cyborg-agents who control them. That means danger according to their definition, and they eliminate danger as something no one could ever desire.

Danger lurks at airports, too. You know at all times that this is a dangerous place as all the visible safety procedures convincingly indicate. Then there are the invisible security checks. Much of it is automated in this cybernetic world. You cannot ask questions, you must not inquire, you must not know, you must obey, and most of all, do not try to joke because that is too human. The dark side rules. Do not test their security systems and its procedures; the robots would not understand, as they have no sense or desire for compassion or empathy. You sense danger but you are safe. It does not matter that you do not want this. At least you can go home and take a bus next time. Cyborgs do not do so.

8. Conclusion

Where are we now? My point is that in a frame like an airport we live together with robots, our cybernetic lives. The desires we have and should have are the desires fit for the technological, automated robot frame. This frame constricts and castrates our desires making them look like a shadow of what might have been. In a paradoxical manner, robots have no desires and they will never have them, but our human desires within the technological frame start approximating the desires robots would have if they had desires. This is uncanny. Are we undead or alive? Perhaps our only hope is rebellion, art, imagination, and radically new designs—praxiology in the new key. Technological design should not suffocate us like that greatest of all primitive robots, the Internet with all its digital tentacles, threatens to do. As Heidegger says, of course, the essence of these new designs is nothing technological. Perhaps a great robot war will descend upon us. This is what C-12 came to tell us.[19]

Notes

1. "Motherboard," in *The Secret Theater*, Queens NYC, October 2012, by AntiMatter Collective. Cf. D. G. Stork, ed., *Hal's Legacy*, MIT Press, 1997 about the killing of Hal.
2. Lydia H. Liu, *The Freudian Robot*, Chicago University Press, 2010, 139 ff.
3. See R. A. Brooks, *Robot*, Allen Lane, 2002, and H. Moravec, *Robot*, OUP, 1999.
4. See G. Thomson, *Needs*, Routledge, 1987, Introduction.
5. See T. Schroeder, *Three Faces of Desire*, OUP, 2004.
6. M. Kirwan, *Discovering Girard*, Cowley, 2005, Ch. I.
7. A. Lemaire, *Jacques Lacan*, Routledge, 1977, 163.
8. Robert M. Geraci, "Apocalyptic AI: Religion and the Promise of Artificial Intelligence," *Journal of the American Academy of Religion*, vol. 76, 2008, 138–66.
9. See my "New Life Forms: Min or Max Cyborgs," *Cutting Through the Surface of Bioethics*, T. Takala et al, eds., Rodopi, 2010, 231–46.
10. About the Soviet attempt to do this, see S. Gerovitch, *From Newspeak to Cyberspeak*, MIT Press, 2002.
11. M. Heidegger, "The Question concerning Technology," numerous translations, see e.g. http://72.52.202.216/~fenderse/Technology.html (accessed October 10, 2012).
12. Liu, 2010, 214.
13. Liu, 2010, 224 ff.
14. See G. Wood, Living Dolls, Faber and Faber, 2002, and E. C. Wilson, *The Melancholy Android*, SUNY Press, 2006.
15. Liu, 2010, 258.

16. Cf. a poem by W. S. Merwin, "Neither Here nor There: An airport is nowhere . . . ," The New Yorker, October 15, 2012.
17. Henry Miller would have dubbed it *The Air Conditioned Nightmare* (1945) (numerous editions).
18. Tom Hanks in the Steven Spielberg's film *The Terminal* (2004) tries to show what it is like to live at airport lobbies, not convincingly.
19. My thanks are due to Professor Heta Gylling and Mr. Karri Liikkanen (Helsinki). My research was supported by TEKES (Finland).

References

Airaksinen, T. "New Life Forms: Min or Max Cyborgs." In *Cutting Through the Surface of Bioethics*, ed. Takala, T. et al., 231–46. Amsterdam: Rodopi, 2010.
Brooks, R. A. *Robot: The Future of Flesh and Machine*. New York: Allen Lane, 2002
Geraci, R. M. "Apocalyptic AI: Religion and the Promise of Artificial Intelligence." *Journal of the American Academy of Religion* 76 (2008): 138–66.
Gerovitch, S. *From Newspeak to Cyberspeak: A History of Soviet Cybernetics*. Cambridge, MA: MIT Press, 2002.
Heidegger, M. *The Question Concerning Technology.* 2012, numerous translations, see e.g. http://72.52.202.216/~fenderse/Technology.html.
Kirwan, M. *Discovering Girard.* Cambridge, MA: Cowley, 2005.
Lemaire, A. *Jacques Lacan*. Boston: Routledge, 1977.
Liu, L. H. *The Freudian Robot: Digital Media and the Future of the Unconscious*. Chicago, IL: Chicago University Press, 2010.
Merwin, W. S. *Neither Here nor There*. The New Yorker, 2012.
Miller, H. *The Air Conditioned Nightmare.* 1945, numerous editions.
Moravec, H. *Robot: Mere Machine to Transcendent Mind.* New York: Oxford University Press, 1999.
Schroeder, T., *Three Faces of Desire*. New York: Oxford University Press, 2004.
Thomson, G. *Needs*. New York: Routledge, 1987.
Wilson, E. C. *The Melancholy Android.* Albany, GA: SUNY Press, 2006.
Wood, G. *Living Dolls: A Magical History of the Quest for Mechanical Life*. London: Faber and Faber, 2002.

From Good Robots to Useful Intelligent Agents

Grzegorz J. Nalepa
Institute of Philosophy
Jagiellonian University
Cracow, Poland

1. Introduction

The concept of robots has been one of the central ones for the artificial intelligence (AI) (Russell and Norvig 2009). In AI, robots were originally associated with machines assisting humans in their everyday life. Robots were supposed to support, help, or protect humans. These machines should be able to do what man cannot, excel in areas where humans reached their limits, or where certain actions would put their life or health in danger. Robots were expected to be more or less universal in terms of scope of actions they would be able to carry out. Thus, implicitly robots would be envisaged to be man-like. So robots should be mobile in order to follow man, or to even substitute man in some situations. Robots should also communicate in a man-like fashion, that is, should be able to speak and interpret voice commands. Besides the practical usefulness of such features, this would make it easier for people to interact with robots on a daily basis. Robots could be machines extending human capabilities, but in some cases they would be expected to be companions to humans, for example, robotic pets. This would make robots in human environment more than just machines (devices). In fact, they could be implicitly treated as living beings. Originally, such a perspective would lead to a number of important questions concerning the status of robots in human society, for example, would robots feel, think, and ultimately would they have consciousness.

Today, decades after the first ideas about robots as a possible future technology, we find ourselves in a society filled with ubiquitous and miniaturized devices that perform semi-intelligent actions. They mostly independently communicate using wireless networks, and processing large quantities of information for us and about us. They are interconnected by invisible yet omnipresent information infrastructure that allows them to exchange information. How far is it from the visions of robotics as a field that was supposed to produce omnipresent assistants or companions to humans? In fact, robotics is a mature and independent field of study and engineering. While the task of producing mobile robots that assist man is still an important and pursued goal, it seems that designers mostly focus on what is practically useful. Therefore, we have stationary robots in manufacturing plans, mobile, remotely controlled or semi independent mobile robots for rescue missions and police operations. One of the original objectives of delivering mobile, humanoid robots is still being worked on. However, these efforts—while fascinating and ambitious—seem to be of limited scope. Designing mobile robots that would be actually suitable for a mass consumer market still seems to be out of reach of the industry. One of the first, and so far few examples are robotic vacuum cleaners. These are the devices that could in fact be the practical beginning of domestic robots.

Delivering a complete robotic device requires practical combination of several factors and requires deep knowledge from multiple fields. In fact, what seems to matter the most is the "brain" of the robot, that is, the component that makes a robot "intelligent." This feature is indispensable for a robot that is expected to perform actions in environment, or to communicate with humans. So it seems that even if robotics became a practically separate domain from AI, it somehow still depends on its results and achievements. AI provides means for modeling the intelligence of the robot on the conceptual level. Therefore, intelligent or sometimes just "smart" robot would often be considered "good" if it were useful, that is, capable to carry out tasks it was expected to do in a feasible way.

2. Intelligent Systems

In AI, the question of what makes an artificial system "intelligent" remains unsolved. I. S. Torsun states that intelligence is easier to recognize than to be defined, and, in fact, AI is not about the search of such a definition and can get along with one (Torsun 1995). However, the lack of answer does not stop the AI researchers from developing new generations of systems that can be called "intelligent" in some way. A helpful tool for categorizing AI approaches to intelligence might be the classification presented in (Russell and Norvig 2009). Four groups of systems are considered there, including systems that (1) think like man, (2) think rationally, (3) act like man, and (4) act rationally. While this is a general perspective on possibly all AI systems, it seems to be quite useful for approaching diverse efforts to develop robots—at least from a popular, not scientific perspective. From the practical point of view, most utility robots, including domestic ones, would be expected to act rationally. So a vacuum cleaner is expected to act (i.e., clean) in a rational way, not necessarily in a human-like one. In this context "rational" mostly means that it is able to carry out actions needed to achieve a planned goal, by finding means needed to do so, and using the available resources in an optimal (in a defined sense) way. To solve certain tasks, for example, planning an optimal path a robot would need to think rationally, where thinking would be logical reasoning—a task that can be automated and programmed. However, what would matter most for people expecting a robot to be their companion would probably require a human-like behavior (e.g., a robot could be unpredictable to some degree) and human-like thinking. The latter is commonly considered to include—in a very broad sense—emotions.

Therefore, here we will mostly focus on intelligent systems that are the brain of a robot, not the robots themselves. Moreover, we will omit the term "artificial," since here we assume the exclusion of any living organisms being used to model the intelligence of robotics systems, even though the work on cyborgs is conceptually possible. Moreover in today's AI artifacts, the hardware plays minor role from the software that runs them. While hardware may be important, sometimes critical to the interaction with the user, it is the software that makes the hardware operate. It is also the software part that captures the "intelligence" of the system. This is why here we will mostly focus on concepts underlying the software part of intelligent systems.

Finally, it stems from the above, that so far, a "good" intelligent system is a system that can be considered useful in some sense. It mostly means that a system or a robot controlled by it can carry out certain action, perform on a task. A task can be "cleaning a room," but it could also be "making a user happy, feeling good, not lonely." It is also implied the "goodness" of a robot is a certain quality that can be assessed by its user. Moreover, it can be designed by an engineer during the creation of a robot, or in a more general case the component that makes it intelligent. Here we will shorty introduce approaches and methods that can be used to design good intelligent agents or systems.

3. Symbolic Representation of Knowledge

One of the principal issues in the classic engineering of intelligent systems is an appropriate structured symbolic system that can be used to represent the knowledge about the domain in which the system operates. This is the so-called knowledge representation method. It should be feasible for the task, providing a language with clear syntax and well-defined semantics. Moreover, knowledge captured using such a method should be easy to be processed in an automated way in the reasoning process (Brachman and Levesque 2004).

Rules are one of the most common and intuitive ways of expressing human knowledge. They capture the knowledge about regularities or patterns in human environments. They are mostly formulated in the processes of abstraction and induction based on observations. Using rules, humans perform deductive reasoning, making decisions based on available facts. Moreover, we often perform abductive reasoning, where specific goals and objectives need to be obtained or satisfied. Both of these reasoning techniques can be easily modeled and executed by intelligent agents. This makes their reasoning schemas close to the human ones. Rules themselves are close to the conditional statements in the natural language having a common form of the "if-then" patterns. Thanks to these features, rules became a widespread knowledge representation and reasoning method for intelligent agents. They are used where the human knowledge is easily available and well conceptualized. In fact, rules are a foundation for some of the most successful class of AI, called expert systems. These systems use general rule-based reasoning with a domain-specific knowledge encoded with rules. They provide relevant and domain-specific solutions for the problems stated by user (Liebowitz 1998).

Building rule-based systems requires certain skills. A good rule base should be complete, thus not missing any rules relevant for solving problems in the specific domain. Moreover, it should be deterministic. Finally it should be minimal, thus non redundant for the agent to operate efficiently (Ligęza 2006). Designing such systems is not an easy task. It often requires a formal model for representation and verification. However, such an approach improves trust a user has in such systems.

For creating intelligent systems that process knowledge in a human-friendly fashion, the issue of the semantic knowledge processing needs to be addressed. What we expect from such an agent is what we would call "an understanding." The motto of the Semantic Web initiative was "I know what you mean!" In fact, from the practical point of view, what we are mostly interested in is an interaction with the agent. We are not concerned whether it has a genuine understanding of what we want from it, or even if it is capable of such. What matters most are its actions conforming to our expectations. Therefore, a good agent behaves in a way that we expect it to do. To deliver such semantic agent, a number of Semantic Web methods were developed (Hitzler et al. 2009). They include the standardization on the syntactic level of the data format the agents use to communicate. This is the so-called XML language. Using it, a format to express basic semantic annotations (meta data) called Resource Description Framework (RDF) was proposed. These basic solutions provide a foundation for the formal ontologies that are the core of the semantic agent knowledge base. In this sense, a formal ontology is a specification of conceptualization. It is a formal model that captures the relationships between concepts used by an agent. Ontologies can have numerous forms and can be used by agent in different ways. In the most common case, an ontology provides a dictionary in the form of a taxonomy of concepts described in the OWL language. Such a solution allows for a definite specification of agent language by restricting the conceptual framework only to concepts that are actually needed.

A practical challenge for ontology-based agents is their communication and knowledge sharing. Agents created by different designers and for specific users can have similar, but not identical ontologies. Thus the problem of ontology matching becomes apparent. In such a case agents need to use specific methods to align their perspectives of the world they act in to establish a common vocabulary that allows them to interact. Such an alignment is non-trivial and requires machine learning methods that allow for discovering concepts meaningful to several agents. Finally, a question arises how it is possible to provide a useful and expressive ontology for an agent, and is such a knowledge engineering task feasible for users themselves? Skillful knowledge engineers can build ontologies dedicated for specific domain, for example, for a medical agent that helps diagnosing specific illness. However, in a general case such a priori creation is not always possible or feasible. In some cases, especially distributed communities of users and their agents, ontologies are created in a semi-automated bottom-up way forming the so-called folksonomies.

In terms of complexity of relations between concepts folksonomies can be considered simple. They are mostly built on observations which concepts are used most often by a specific community and how they are used. An ontological knowledge base mostly represents a static structure of a conceptual framework. Using this structure, an agent can perform the reasoning process to comply with user needs and expectations. Such a reasoning is most often rule-based (Nalepa 2011).

In classic knowledge engineering, human knowledge is being acquired and represented, forming knowledge bases. However, a practical realization of such a process can encounter a number of practical difficulties. These may include lack of cooperation from human experts and problems in expressing what they know in a way interpretable for knowledge engineers. This is why an approach complementary to the manual creation of knowledge bases includes the use of machine learning techniques. These enable the system to be trained, or even learn autonomously in the situation where human experts are not available, or their knowledge is incomplete.

4. Agents that Learn

Learning has always been an important aspect of intelligence in AI systems. In fact, the ability to learn has always been considered one of the distinct features related to the intelligence itself. However, in last twenty years, the domain of machine learning rapidly developed and evolved. Currently, it is a mature field, often separated from the core AI itself.

Machine learning can be understood as a "systematic study of algorithms and systems that improve their knowledge or performance with experience" (Flach 2012). Due to the use of machine learning methods systems can be made better suited for the task needed to be carried out. The field itself is quite heterogeneous. However, all machine learning approaches share a common conceptual framework. A machine learning task is some abstract conceptual representation of a problem to be solved with respect to objects in some domain. These objects are described with the use of features used to provide a certain language. The task itself requires, in most cases, making some decision based on the features. Most often this can be the task of classification. The performance of the task is improved during the learning process of the system. In order to make such a decision, a certain formal model is needed.

The approaches to the learning process can be characterized in two main ways. First of all, we may consider a situation, where the agent is taught using certain examples, in such a case the supervised learning process is considered. Where there is no appropriate example or some other prior knowledge, an agent can also learn in an unsupervised fashion. Depending on the outcome of the task, the model to be used may be a predictive one, or a descriptive one. The former uses predefined set of concepts to encode the decision (e.g., classification), where the latter comes down to the creation of concepts suitable for describing the outcome of the process.

Finally, diverse models for the learning tasks were proposed. They could be generally divided into geometric, logical, and probabilistic ones. Geometric models are based on the concepts from geometry, such as distance, lines, or planes. An example of a geometric model is regression. Logical models describe the features with logical formulas that can be most commonly represented in the form of rules or decision trees. Finally, probabilistic models use the probability distributions to predict optimal decisions. A common probabilistic model is based on the Bayes theorem. There is a large variety of machine learning methods, but as Flach states "models lend the machine learning field diversity, but tasks and features give it unity" (Flach 2012).

Learning allows not only for the automatic creation of knowledge bases but also for the intelligent systems to improve, thus becoming better, in the tasks they perform. Furthermore, thanks to machine learning methods, systems can adapt to the needs of the user. Ability to adapt can be considered yet another important feature of useful intelligent systems.

5. Adaptable Interfaces

Interaction with an intelligent system requires a good user interface. Such an interface should be intuitive, thus not requiring any additional training or documentation. In many cases, interfaces are based on certain visualizations of system functionality. Visual user interfaces make it easier for a user to use the system, since they use some common visual metaphors. Thanks to them, a system can mimic a well-defined part of the user environment, for example, a desk, library, and so forth. In fact, visualization is not only important on the interface level but visual methods of designing the knowledge possessed by the system are also commonly used. Ontologies can be visualized as semantic networks composed of nodes representing concepts and arc representing relations. Rule sets can be represented as decision tables or decision trees that correspond to the control logic. Such visualization allows the user to better understand how the systems are built.

Recently user interfaces became multi modal, combing classic point-and-select solutions with visual and voice commands, as well as touch-based interfaces. Moreover, the use of many sensors in the interface devices allows for capturing a handful of additional information, for example, where the user or the device is located. Newest proposals include the ability of the user to treat certain artifact as components of interfaces. The so-called tangible interfaces aim at reusing common objects present in the user environment, for example, a pen, as interface. Furthermore, an interface should not be static. Recently it has become apparent that interfaces should be adaptable. Instead of having the user to learn how to use an interface to properly interact with the system, its interface should be able to learn an optimal way for the user to use it. This can be achieved by rearranging certain components of visual scene of the interface. The system should also track user interactions to capture and correct its mistakes during the interaction. Such interfaces are in fact self-optimizing.

An ultimate goal for many interfaces is to provide the so-called augmented reality environment. In such a setup, perceptual data provided to the user is altered, augmented with a system-generated content that enhances the perceptual experience. Using augmented reality interfaces, intelligent systems can enrich the sounds the user hears, and images he or she sees. They are built based on glasses or headphones that filter the image or sound. The system can add additional, context-dependent information providing more information to the user. This in turn allows for specific perceptual data fusion and increases the amount of information the user gets from the system. Current augmented reality interfaces include glasses integrated with dedicated head-up displays.

6. Distributed Systems

An intrinsic feature of today's intelligent agents is their network interconnection. Most of the systems are equipped with multiple communication interfaces that can be used to exchange information in the local environment or on the global scale using the Internet infrastructure. The Internet itself became a planet-wide network that combines most of the computer systems. While according to the original design, most of these systems where servers provide data to clients, the current trend is to move to a more unified situation, where peer-to-peer networking allows any participant of the communication to be both consumer and producer of information.

Another important paradigm shift concerns the level of storing and processing of data. A new level of communication is considered, where the web of data is being replaced with web of documents. These documents are to be structured and annotated with the previously mentioned Semantic Web methods. They are also heavily linked with other documents, often published by public and governmental bodies and organization, according to the Linked Open Data policy. This allows leveraging the communication from the content level to the knowledge level, where structured content is semantically annotated, put in a context, and related to other relevant information available on the web.

These changes make many of today's intelligent systems distributed. They no longer use their own, internal data banks, or knowledge bases. On the contrary, they are often designed to harvest information automatically found on the web, and combine it to perform specific actions. There are many different architectures of such systems considered within the theory the so-called multi agent systems. A good multi agent system should be able to coefficiently communicate asynchronously with others in a heavily distributed and possibly unpredictable environment using the specific communication protocols. In some cases, an agent should not only be able to interact but also to compete, for example, to gain information. Such behavior is often modeled using the game theoretic approach (Wooldridge 2009).

What makes network-centered systems even more complicated is the separation of hardware and software parts by means of virtualization. The software that implements the logic of the intelligent systems no longer needs to be executed on a given machine, hardware, that is, computer system. In fact, it is becoming very common to provide a virtual hardware environment that perfectly mimics the physical one. A software system then runs in a virtual hardware system that can be easily migrated between physical hardware systems. So a single software agent no longer needs a specific machine to run. Moreover, it can easily move between physical locations. Furthermore, the execution of the code of such an agent can be distributed to several systems in different physical locations that run in parallel. In such cases, a so-called software cloud is considered. It is a purely virtual software environment, based on a network of distributed and heavily interconnected hardware systems that provide a complete ecosystem for intelligent agents that can move freely. This makes such an agent free of physical location.

To complete the picture of the massively networked environment, the software cloud in which agents run is integrated with the network of omnipresent ubiquitous mobile devices used by humans in the physical world. These devices allow feeding any information in the cloud, as well as acquiring context relevant data to humans using them. This pervasive computing environment makes it possible for the humans to access instantly and in any location almost any information from the cloud. At the same time the devices feeding information about human activity to the cloud make it possible for the cloud-based agent to track humans to support and assist them. Such a system greatly leverages the comfort a human user has with respect to the information and service access. However, it raises some grave privacy concerns and questions of the impact of the increasingly intelligent systems of our everyday life.

7. Context Awareness

Having knowledge available almost anywhere and instantly is a fascinating opportunity. However, it is also a grave challenge to find the relevant and valuable knowledge. It is not only the problem for humans but also for the intelligent agents. This is where the concept of context awareness comes in. The exposure of context is related to the fact that representing the possibly complete world knowledge in any situation is so far not possible. At the same time, human intelligence is often about recognition of what knowledge is relevant in a given situation of state. Thus, context-aware systems need to recognize such a context, select relevant knowledge to be able to take or recommend relevant action. According to Anind Kumar Dey, context is "any information that can be used to characterize the situation of an entity" (Dey 2000). This information may be location of the user (spatial context), presence or absence of other devices and users nearby, or collaboration with other users (social context), time (temporal context), user behavior or activity, and possibly any other environmental data gathered by microphones, light sensors, and so forth.

Contextual data can be provided by the pervasive computing environment, and can include data acquired from mobile devices, for example, human location, along with his or her physical condition. Once fed into the cloud, it is used by the cloud-based services, coupled with agents running in the mobile devices to provide the human information needed in a given situation. Such architecture can use machine learning methods to gain knowledge about patterns of human behavior, including his or her daily habits as well as routines changes in the environment. Therefore, an intelligent system can learn that a person wakes up in the morning, travels to work at certain times by certain means, recognizes other people that are met on the daily basis, and so forth. In such architectures, systems gather contextual data and learn to classify them to reason about human activities.

This allows an intelligent system to actually predict the human needs, for example, knowing the daily routine of a human, an agent in its mobile phone can notify the agent in the office that the person is approaching, and knowing the fact that he did not have time to drink a coffee in the morning, thus asks the coffee machine in the office to prepare one.

Context processing technologies make intelligent systems context aware. This awareness includes the ability to learn, classify, and recognize context, and more importantly reason about it. It is a foundation for the so-called smart-environments, composed of omnipresent interconnected, and possibly mobile devices using the services provided by the cloud of agent to assist the human and automatically adapt its environment in a given situation, such as home, office, or even a more public space, to its needs, habits, and expectations. Again, while it leverages a certain comfort humans have, it makes them more and more dependent of the agent architecture present in the background.

8. Conclusions and Outlook

Here we considered how the classic vision of robots can be related to today's intelligent systems—in a broad sense. What was once expected to be an independent mobile human-like device, or even being, today is often a system composed of diverse hardware suited for its specific function coupled with special software part (a program in a general sense) that makes it behave—operate in a technical sense—in an "intelligent way." By intelligent, we mostly mean rational, that is, capable of carrying out its goals. When such a system becomes autonomous and handles interactions with environment, it can be called an intelligent agent. Such intelligent agents are omnipresent but need not to be embodied.

To create such agents, AI developed a number of methods and techniques. Agents use symbolic knowledge representation to describe knowledge about the environment and reason about it. To automatically create or extend such knowledge bases they use machine learning techniques. Furthermore, to optimally interact with humans and the environment they have adaptable interfaces. Such agents rarely are separated. Most often, they are heavily interconnected, and distributed using computer networks, storing information in data clouds. To discover and use relevant information they are context-aware to adapt to a particular situation. These general features that can make an intelligent agent useful, and from practical point of view serve well its goals, is termed as "good."

The techniques discussed here are not the only ones. Many AI approaches use the so-called computational intelligence where the behavior of an agent is modeled using numerical approaches, often nature inspired. There is also an important development of probabilistic models, including the field of probabilistic robotics. From the study of distributed networked models the concept of collective intelligence also emerged. Finally, there is a growing interest in cognitive approaches, including cognitive architectures for intelligent agents, and cognitive robotics, which offer a more holistic view of the domain.

So what would be the landscape of AI, robots, and intelligent agents in the future? Probably different to most visions we have now. Rapid development of hardware, including radio networking technologies, and new computational models can very fast change this landscape. The development is driven by the technology itself, and by user needs, which are often of short term. The systems we build become more and more intelligent, thus, useful. At the same time the machine intelligence does not seem to be much similar to the human one, which in a way still remains a mystery. Apparently, to complete the picture the research on rational systems might also be extended in the directions of machine emotions and computational creativity. This could lead to the creation of machines that are even more human like. However, the practical usefulness of these agents would still remain an important factor.

References

Brachman, R., and H. Levesque. *Knowledge Representation and Reasoning*. Boston: Morgan Kaufmann, 2004.

Dey, Anind Kumar. "Providing Architectural Support for Building Context-Aware Applications." Georgia Institute of Technology, 2000.

Flach, P. *Machine Learning: The Art and Science of Algorithms that Make Sense of Data.* Cambridge: Cambridge University Press, 2012.

Hitzler, P., M. Kroetzsch, and S. Rudolph. *Foundations of Semantic Web Technologies*. Boca Raton: Chapman & Hall/CRC, 2009.

Liebowitz, J. *The Handbook of Applied Expert Systems*. Boca Raton: CRC Press, 1998.

Ligęza, A. *Logical Foundations for Rule-Based Systems*. Berlin, Heidelberg: Springer-Verlag, 2006.

Nalepa, G. J. *Semantic Knowledge Engineering a Rule-Based Approach*. Krakow: Wydawnictwa AGH, 2011.

Russell, S., and P. Norvig. *Artificial Intelligence: a Modern Approach*. Englewood Cliffs, NJ: Prentice-Hall, 2009.

Torsun, I. S. *Foundations of Intelligent Knowledge-Based Systems*. London, San Diego, New York, Boston, Sydney, Tokyo, Toronto: Academic Press, 1995.

Wooldridge, M. *An Introduction to MultiAgent Systems—Second Edition.* John Wiley & Sons, 2009.

On the Significance of "Responsibility" in Technological Thinking

Ladislav Tondl
Institute of Philosophy
Academy of Sciences of the Czech Republic
Prague

1. Responsibility and the World of Technology

Responsibility is traditionally understood in terms of values, primarily as a duty, stimulated and conditioned ethically, morally, or through other sources of human values, of the subjects of rational and thus goal-directed action aimed at selecting humanly acceptable objectives and means of their action, while paying heed to its results, including potential future impacts. An important place in the broadly based realm of goal-directed actions is occupied by those types of actions in which human creations are made and, hence, different areas of artifacts, including technical artifacts which substantially change the content and means of human self-fulfillment, while improving and enriching this content and its range, and, at the same time, posing new dangers and new risks. It was the Polish philosopher T. Kotarbiński, who, as one of the leading figures of a world philosophical congress to be held in Vienna in 1968 called on the author of this study to prepare an invited paper on the key issue of the philosophy of technology. (The paper was published in German under the title "Januskopf der Technik," that is, the dual face of technology, and anticipated a novel theme in contemporary technical thinking, most frequently described as "technology assessment.") This particular mode of thinking and reasoning, that is, an ability to perceive and respect not only the positive aspects, benefits, and advantages of newly created as well as applied artifacts but also potential accompanying negative impacts or risks, calls for the necessity of integrating knowledge of different nature and different provenances but also for the need to integrate knowledge and values. Also seen in this sense, the legacy and traditions of Kotarbiński´s well-known treatise on efficient action should be taken as a major challenge posed also to the civilization's new trends and programs, such as images of the information society, knowledge society, or society capable of technologically implementing or supplementing even very challenging and sophisticated human action.

A major role in the sequence of changes of our civilization is played by communication processes and the interconnected information technologies. Changes associated with the spread of written communication, book-printing, an ability to read the press, listen to the radio, watch television, use the Internet and other areas of communication and information technologies have not only changed the content and direction and the value atmosphere but have also taken a lion's share in generating what may be characterized as images or social patterns of the desirable, preferred world or teleological world. Not even the contemporary globalized society can exist without specific visions or images of those possible worlds which are more perfect, advantageous, and more pleasant to live in. During consideration, critical decision-making and possible selections of the ever more expanding spectrum of various alternatives and options we can hardly eschew the questions relating to this particular approach, associated with the principle or concept of "responsibility."

In addition to information technologies ensuring predominantly communication processes and communication functions, technological artifacts and their systems, which have primarily managing and decision-making functions to play and thus the task of automating those actions that were traditionally ascribed solely to the human intellect, have now been gaining ground to an ever greater extent. These are mostly spheres of covering artificial intelligence, technical artifacts, and their systems, utilizing not only empirical findings but also well-known and confirmed generalizations and rules, including inference rules.

2. Major Starting Points

In present-day discourses, debates, and other important messages and statements the word "responsibility" has grown to be one of the most frequently used terms, usually accompanied by different attributes (such as personal, social, political, economic, moral, legal, or other kinds of responsibility), and the same is true of diverse personal and social discussions, public speeches, or even arguments. Moreover, this word and the term associated therewith may evoke very different and sometimes even contradictory contexts. The use of the word "responsibility" may conjure up allusions to merits or abilities of subjects as well as point to someone's guilt; it may involve praise lavished on an author or initiator, just as, in some cases, his criticism, accusation, or even reprobation. This word, especially when used in expressed or published speeches, has become quite a popular term, usually implying an unspoken assumption that all the recipients understand it well and identically. The term "responsibility" is suggestive of connections with that form of communication that comprises answers to uttered questions and queries, or at least wishes or sentiments aimed at learning, ascertaining, or knowing something. Unspoken or, to be more precise, usually unexpressed questions may relate to various kinds of queries, for instance questions evoked by various question operators, such as the expressions "who," "why," "for what purpose," or other possible questions connected with goal-directed action, with its intentions, anticipations, means, conditions, and other circumstances.

One of the major connections is the relation between responsibility and goal-directed action, whose results are for certain artifacts, that is, for artifacts of material, technical as well as intellectual nature—their level, quality, and other possible value-related dimensions. In a similar vein, one can consider responsibility of a subject of specific action in the light of the goals, anticipated results of the action concerned, or in view of possible losses and risks of such an action. At the same time, it is evident that responsibility constitutes a relationship or function with several possible arguments or, to put it differently, it is invariably responsibility of someone, for something, in view of possible results, impacts, or other possible contexts. Also, this is usually expressed in a way that the term responsibility is related to a specific subject, to a specific theme, or possible contexts. These reflections imply that the significance of this term and the interrelated concept can be best clarified in the light of a specific conceptual framework which is applicable to the process of communicating a description, recording or explanation of an event, decision-making, or selection of possible alternatives within a certain domain or thematic field.

3. Responsibility and Goal-Directed Action

A major conceptual framework for the semantics of the term "responsibility" is provided by various spheres of goal-directed and thus rational actions, structures of such actions, including primary decision-making, target orientation, applicable means, attained results, and their possible impacts. This is clearly illustrated by the oldest forms or patterns of such actions. Such were the efforts for taking care of human health in ancient Greece, based on a set of rules of ethical nature, that is, the Hippocratic Oath, endeavors to maintain physical fitness proceeding from the tradition of sporting competitions and Olympiads. One cannot ignore the significance of what can be summarized as accents placed on the spiritual, intellectual, knowledge, and value-related prerequisites of those human actions, expected to result in a human construct, in an artifact of material, technical as well as spiritual nature, which originates first as a purport, an intention, gradually changing into a plan, a project, or a program comprising follow-up partial actions, and ending in a final result, a finished work, and hence an artifact, reflecting the original purpose, mental image of a given target orientation. Therefore, it does not hold that a newly created artifact is primarily a product of man's creative action; this particular action should be preceded by a wish, an ambition, an intention, and a project that can also be interpreted not only as an information model of a future artifact but also as a program, an organization, and an arrangement of a sequence and integration of individual follow-up actions. That is why the principles, content, and requirements of human responsibility are also related to those ideological, intellectual, knowledge, and value-related prerequisites of a goal-directed action which is expected to lead to a desirable act, situation or a new work supposed to be the outcome of a given action. (In this respect, some authors speak of the ethics according to conscience or the ethics of responsibility, which is spelt out, with considerable accent, in those parts of the New Testament traditionally described as the "Mountain Sermons" or "Sermons on the Mount.")

Responsibility of all actual and potential subjects of action, and thus not only of the subjects of initiatives and suggestions, authors of whole components or stages of design activity, actual organizers and supporters of goals, available resources and means of actions, their genuine and possible results comprises their evaluation, including assessment of their possible and acceptable alternatives, their selection as well as practical application. In these contexts, authors usually accentuate the fact that choice of specific goals cannot and should not be justified by any means whatsoever, and that it, therefore, does not and should not hold that "the end justifies the means." Responsibility of the subjects of goal-directed actions is known to have significant relations and importance vis-à-vis the time dimensions of such actions, and especially their results. Standing in the centre of attention in such contexts is usually the duration of such reliability and the interconnected possible guarantees, time guarantees, and their fixed or guaranteed limits.

It is undoubtedly an important value of the traditions of some professional or specialized activities that the ability of a given subject of such actions to carry and guarantee responsibility for quality performance of such an action and its results had to, and often was obliged to, be proved beforehand, particularly before the start of such work. Responsibility for a specific action and its anticipated and potential results is always dependent on the qualification, competence, and abilities or skills of the subject of action, and thus of the author, proponent, initiator, designer, and implementing subject of a planned or intended work just as of its user. That is why a good deal of great importance is ascribed in such contexts to qualified preparation of a given project, complete with preparation for its practical execution and subsequent application, adequate awareness of potential risks, their prevention even in possible and practicable solutions. In some spheres of demanding or qualified actions, it has been a matter of course to check those prerequisites, or rather to test the specialized and qualification prerequisites within specific time intervals.

In many areas of important and particularly demanding actions, the process of securing a sufficient level of responsibility was traditionally evaluated and verified by means of qualification yardsticks, by documents attesting to attained education or abilities verbally to reproduce the necessary knowledge, working procedures, or applicable methods. Current experience, particularly in the sector of social and political work and relations, in the field of social and economic decision-making, shows that the above prerequisites, primarily the qualification requirements, and even utter neglect of such requirements may open the door to various negative phenomena and maladies of contemporary society, whether we call them a chase to get rich quick, an unnatural yearning for power and money, corruption, clientelism, and other pernicious and particularly contagious diseases of the present-day society.

In addition to adequate qualification, relevant knowledge, and other prerequisites for utilizing and applying knowledge, the following aspects are known to contribute to responsibility of a subject of action, while frequently and also markedly shaping the subject:

- value prerequisites and value orientation of the subject of given action, its moral and ethical standards,

- its ability to consider potential risks and dangers of undesirable abuses, to take into account broader contexts of an intended procedure in the start-up and implementation of the actions planned,
- an ability to take into account and to respect other opinions, other procedures, or alternatives and options under consideration,
- communication prerequisites, for instance prerequisites for conducting broader interdisciplinary consultations and negotiations.

4. Responsibility and Meritorious Questions

The above remark involving connections between responsibility and questions or queries has indicated that it is, indeed, possible to ask questions and enquire, primarily about the nature of specific responsibility, subject of responsibility and its share, nature of a given action, both intended, prepared or just considered one, to find and demand answers on diverse topics, demand explanations and justifications of proposed or already implemented steps, and their outcomes and impacts. These may be different forms of questions, especially queries involving operators such as "who," "what," "why," "how," "what for," and many other questions including the traditional "cui bono," "cui prodest," that is, to whose benefit? Generally speaking, we should stress the usefulness of humans and primarily civic inquisitiveness, a quality that should form a permanent component and a major criterion of positive life of a civil society, and hence an opposite to mortifying lethargy and indifference.

It is natural that standing in the centre of attention should be questions pertaining to the subject, and hence carrier, of any kind of responsibility. Traditionally, under public eye and usually also under scrutiny is the qualification of a given subject, its qualification standards, erudition, and know-how attained during the period being monitored. (At the same time, success in previous work, former positive results are no full guarantee of achievements in a new sphere of action.) It is quite natural that some demanding and highly specialist work and the professions associated therewith may be carried out by people with demonstrably high professional competence, and that this particular qualification is or should be checked by necessary forms of certification and accreditation. For some work and specializations, an important component of what is called global competence is also formed by value-related and ethical requirements, what is usually described as moral integrity, an absolutely clean personal and criminal record both in contemporary life and in previous work. To date, the task of monitoring and providing adequate value-related prerequisites has not yet been sufficiently secured for some types of professions and some executive and managerial jobs where certain subjects, who have acquired powers to handle enormous wealth and property belonging to others, especially the property of individual communities, large groups of people or public property in general, have been known to abuse their positions to siphon off the property in a bid to get rich soon and easily. Well-known experience dating to the past period has also shown that the door to such opportunities has been opened by work in the political sector, including the possibility of using and abusing public property, public orders, membership of executive and auditing bodies, positions in representative authorities and institutions. Also in these areas it is vital to check and monitor responsibility on a long-term basis, keeping an eye on the connections and interlinks between power and considerable wealth and, at least, asks questions who is responsible, who is in charge of what, and how to combat and eliminate irresponsibility.

Reviewing the list of major questions relating to the subject of responsibility, we come to the conclusion that there are two different orientations or directions of queries which are undoubtedly quite substantive, namely:

- orientation seeking a subject's actual intentions or, to put it in terms of a well-known philosophical school, intentionality of questions,
- orientation seeking a subject's decisive goals, particularly the desirable, intended, or anticipated goals of its endeavors.

Human efforts, strivings and quest, as projected into thinking and reasoning, very often oscillate between two worlds, the actual world surrounding man, in which he lives, in which he asserts himself, and out of which he cannot extricate himself, and the world of his desires and aspirations, a desirable world, or—to put it in other words—a teleological world. That is why people come up with wishes, needs, and topics to change, improve, or newly create something in the actual world, hence intentions and plans which can easily be transformed into initiatives, visions, or programs; that is why people have images and illusions of a more perfect world, plans and gradually also projects of objects, facilities which can improve things and conditions prevailing in the actual world. Seen in this light, it is only natural that this may give rise to differences and problems in distinguishing the illusory and the practicable, problems of ensuring such practicability and arguments about the acceptability of the practicable. This also involves the fact that many new solutions, newly created conditions and situations may be connected with risks, threats to human lives, aggravation of the living conditions or the natural and living environment. The paths from the actual world, with its difficulties and shortcomings, to possible worlds offering alleviation of difficulties and drawbacks, greater comfort, visions of perfection and affluence, are not accompanied only by successes, major and wide-ranging accomplishments, and this applies not only to human users but also to the technical equipment used, notably new types of technologies, for example, biotechnologies, information technologies, automated procedures, and the AI. Also the new types of technologies may be "double-faced," that is, bring benefits and enrichment of human life but also pose new risks, new previously unknown and unsuspected dangers. This, however, holds true of all shapes and forms of innovation and innovative changes, hence not only of technological and economic changes but also of organizational, institutional, and socio-political ones. It is definitely a good and beneficial situation when the information media and information technologies are in a position to enrich many people intellectually, cognitively, and culturally, to broaden their horizons or bring them highly valuable cultural values. However, such media and information technologies may also be guilty of manipulating with the opinions and value attitudes of uncritical or inadequately "literate" recipients; they may be guilty of dumbing them down or misleading them, as the media dominated by monopoly power and monopoly ideology had done in the recent past.

Contemporary information media are known to shape very significantly the people's current life, their social relations, target orientations, and choice of applicable ways and means, doing so particularly through the patterns of acting and behavior they present, through the choice of means they use, through the forms and shapes of possible human self-fulfillment and interests they offer. Also in these contexts, we can hardly avoid asking the question to what extent are those patterns beneficial and useful, and hence questions associated with the role played by responsibility, values, and value-related climate. Therefore, it is crucial to make such questions crystal-clear and to put them across quite comprehensibly, especially in the information media and in speeches delivered by those personalities whose statements enjoy public trust and credibility.

As for the fields of information concepts of meritorious questions involving the principle of responsibility, attention is focused primarily on the sector of sources, authors, on securing one's own information space, while the quality, competence, and training of recipients, authors as well as users should not and must not be ignored either. Much more difficult and challenging is how to focus meritorious questions concerning the issue of responsibility in the branch of fully automated processes and thus in the field of robotics and AI. It is precisely these contexts that give rise to the question spelt out by the term "efficient robots." Understandably, the subject of responsibility is and always will be human, namely an individual human subject, operating in all the phases and stages of the generation, decision-making, creation and utilization of a technical work, starting with primary decision, designing and creating, and ending with the possible forms of use and potential abuse of such a work.

5. Major Attributes of Responsibility

In conversations and statements, responsibility is usually characterized and specified in greater detail, frequently by adding different attributes denoting the topics to which the said responsibility is related. Seen in this light, people for example speak of political, legal or moral responsibility, economic or medical responsibility. In such contexts, people then tend to hold discussions and argue about the applicability of different types of responsibility, possible impacts of such kinds of responsibility, for instance whether the so-called political responsibility is enforceable by any other means than political ones, and so forth. That is why it is always advantageous if a given type of precisely delineated responsibility is defined, in no uncertain terms, by appropriate rules or standards. Such rules and standards also make it abundantly clear who can be the carrier of responsibility, which particular requirements and prerequisites must be met, notably in terms of age and health.

Responsibility is principally a personal and individual matter, that is, it is actually attributable to individual subjects, and hence to individual persons. However, in formal terms, responsibility is also borne by the so-called corporate bodies, specific organizations, institutions, and authorities attached to individual social and political groupings. These do and should always have their representatives, their officials, heads, members of their governing boards, supervisory boards who are bearers of the actual responsibility. Formally they are also responsible for the criminal and fundamentalist ideologies, such as the ideologies of totalitarian states known from the past century. Furthermore, it should be noted that after World War II the international court in Nuremberg gave stiff punishments to the individual representatives of the criminal system for their individual shares in the crimes against humanity. That is why it is impossible to condone the principles of collective responsibility, as applied by the totalitarian systems and their ideologies, principles which many people endorsed in the past and to which even some citizens seem to succumb at present.

While importance is attached to the finding that responsibility is an individual and personal property, then it is no less significant for many members of the human community to find and identify a subject endowed with qualities and hence with personalities they can trust and rely on. Even though such aspects of human psyche may describe, undoubtedly with a touch of irony, as the residues of people's tendency to seek the common herd, as a surviving need of having a leader of the herd, these are the feelings and requirements that can still hold masses of people hostage. (Just a personal memory from my childhood in way of illustration: I remember that our young German friends we used to play with in the streets of a South Moravian town in the mid-1930s spoke both "land languages" and suddenly they refused to talk to us, changed their way of clothing and began attending the so-called Fackelzüge, that is, torch marchers chanting nationalist slogans. These memories were brought back to my mind in the 1950s when some of my classmates would meet in closed ranks shouting slogans "Forward with Klement Gottwald (a postwar communist leader in Czechoslovakia) at the helm. . . ." Their opponents would substitute the second part of this chant with the rhyme indecent to be repeated in this text. Unfortunately their words later proved to be true.)

However, contemporary man is a social animal, living, working, carrying out his intentions, and meeting his goals within specific social relations, within certain communities. He can hardly ensure his existence as a Robinson on a desert island or a pilgrim in the wild or deserted wilderness. That is why the goal of selecting and implementing responsibility, and social responsibility in particular, goes hand in hand with a good and well thought-out choice of social relations. In other words, a man in the contemporary social and cultural milieu has to decide and choose to what and to whom he is prepared to give his confidence, with whom he is prepared to cooperate, which particular group of people he is willing to join, whom he will choose as his models, friends or people who will represent him, or to whom he will entrust his voice and representation of his own standpoints and value positions. However, he may sometimes go wrong or be deceived. In such a case, it is prudent and responsible to admit one's mistake. Remarkably enough, the old tradition expressed in various advices and recommendations tends to lay accent on prudence, responsibility, consideration of possible or future impacts, as aptly spelt out by the ancient Latin recommendation: "Quidquid agis, prudenter agas et respice finem."

When demanding enhanced attention and care to be given to securing and especially also to more precisely specifying responsibility, requirements for clear-cut and distinct delineation of the content of responsibility and demands for real and potential subjects of responsibility, we also have to spell out a serious warning against any negation of responsibility, that is, against spreading or merely tolerating irresponsibility. Such a negation of responsibility is often the cause of crises, confusions, and especially losses.

References

Der Januskopf der Technik, 1968, In: Akten des XIV. Kongress für Philosophie. Bd. II, Wien, S. 570–77, German.

Věda, technika a společnost. Soudobé tendence a transformace vzájemných vazeb, 1994, Filosofia, Praha (in Czech, Science, Technology and Society).

Technologické myšlení a usuzování. Kapitoly z filosofie techniky, 1998, Filosofia, Praha (in Czech, German translation, 2003 Sigma eds., Technological Thinking and Reasoning).

Hodnocení a hodnoty, Metodologické rozměry hodnocení, 1999, Filosofia, Praha (in Czech, Values and Evaluation).

Technological Assessment and the Start of Technological Solutions, 1996, In: M. R. Dence, E. Kashner, eds., *Constraints to Freedom of Scholarship and Science*. The Royal Society of Canada.

Problem Areas of the Philosophy of Technology, 2008, In: W. Gasparski, ed. *Praxiology and the Philosophy of Technology*, Transaction Publication, New Brunswick, 25–44.

Znalost a moudrost, 2008, Forum scientia et sapientia, vol. 15, N.1, Bratislava (in Czech, Knowledge and Wisdom).

On the Role of Values in the World of Technology, 2012, In: Olli Loukola and W. Gasparski, eds. *Environmental Political Philosophy*. Transaction Publication, New Brunswick, 179–96.

Notes about the Authors

Timo Airaksinen, (PhD Turku University, Finland 1975) is Professor of Moral and Social Philosophy, University of Helsinki, Finland. He is a Life Member of Clare Hall, Cambridge and Vice-President of the Philosophical Society of Finland and the International Berkeley Society. He is Managing Editor of *Hobbes Studies*. He has written *The Philosophy of the Marquis de Sade* (Routledge 1995), *The Philosophy of H. P. Lovecraft* (Lang 1988), and *The Ethics of Coercion and Authority* (University of Pittsburgh Press 1988). He has also written high school and university textbooks in philosophy and nonreligious moral education. Professor Airaksinen is a member of the International Advisory Board of the Praxiology series within which he coedited *Practical Philosophy and Action Theory* (Transaction 1993) and *Praxiology and the Philosophy of Technology* (Transaction 2008). He is a Life Honorary Member of the Learned Society of Praxiology.

Roberta Annicchiarico, PhD, is a geriatrician. She has been a Medical Assistant at IRCCS Fondazione Santa Lucia, Rome since 2002. She is a contracted professor of Internal Medicine with focus on geriatrics at the school of Physiotherapy, University of Rome, "Tor Vergata" since 2004. She is also a contracted professor of Geriatrics at the School of Nurse, University "Tor Vergata," since 2005. She participates in several European projects. Her main research interests are the disability correlates of chronic diseases and possible clinical applications of new technologies.

Tadayoshi Aoyama is Assistant Professor of Hiroshima University, Higashi-Hiroshima, Japan. Until 2012, he was a PhD student of Nagoya University, Nagoya, Japan. Professor Aoyama has published books and journals in the area of humanoid robotics, bio-inspired robotics. His articles have appeared in the IEEE/ASME Transactions on Mechatronics, Conferences of Robotics and Automation Society, IEEE.

Adam Borkowski had been professor of robotics and applied computer science at the Institute of Fundamental Technological Research, Polish Academy of Sciences (IPPT PAN) in Warsaw until he retired in 2012. In 1978–1979 he was Humboldt scholar at the Universities of Stuttgart and Wurzburg, Germany. In 1990–1991 he was visiting professor at the University of Stuttgart, and in 1994 he worked at George Mason University, Fairfax, Virginia under the support of Kosciuszko Foundation. Professor Borkowski published five books and over hundred papers in the areas of applied mechanics, computer science, and robotics. His books were published by Elsevier and Springer, whereas his articles appeared in Journal of Robotics and Autonomous Systems, International Journal of Automation in Construction, and Lecture Notes in Computer Science.

Jack Brzezinski is an educator, research scientist, and software developer. Currently he teaches programming, robotics, 3D graphics, and mathematics courses. Brzezinski received his MS degrees from Poznan University of Technology and Adam Mickiewicz University. He holds a PhD degree in artificial intelligence (AI) from DePaul University. He has published in the areas of machine learning, data mining, natural language processing, information system management, physical chemistry, and statistical mechanics.

Ulises Cortés is Full Professor and Researcher of the Technical University of Catalonia (UPC) since 1982 (tenured since 1988 and habilitated as Full-Professor since 2006) working on several areas of artificial intelligence in the Software Department including knowledge acquisition for and concept formation in knowledge-based systems, as well as on machine learning and in autonomous intelligent agents. In collaboration with Hospital Santa Lucia (Rome) and University of Malaga since 2001, he has been working in the development and deployment of e-Tools: Assistive technologies for the disabled and for the new generation of senior citizens. Since 2007, Professor Cortes is the Academic Coordinator of the Barcelona Supercomputing Center and the Academic Coordinator for the Interuniversitary Master programme on Artificial Intelligence since 2005.

Toshio Fukuda is Professor of Nagoya University, Nagoya, Japan. Until 1989, he was Associate Professor of the Tokyo University of Science, Tokyo, Japan. Professor Fukuda has published in the area of intelligent robotic systems, cellular robotic systems, mechatronics, and micro-nano robotics. His articles have appeared in the IEEE Transactions,

International Journal of Robotics Research, Advanced Robotics of RSJ, and Conferences of Robotics and Automation Society, IEEE.

Zbigniew J. Gackowski had experience in industry, public administration, and academia. His teaching and research bridge the Central European and US experience in Computer Information Systems (Warsaw University of Technology, The University of Michigan [*Fulbright Research Scholarship*], Purdue University [Visiting Associate Professor], Baruch College [Visiting Professor], CSU, Stanislaus [Visiting and Tenured Professor], and the University of Melbourne [Visiting Professor]). His research had received much recognition. While in Poland, he published more than 120 items, among them four books and six papers in refereed journals and twelve papers presented across Europe, the United States, the Middle East, and South America. While in the USA, he taught at three USA universities and wrote thirty-two research papers that were published in the proceedings of ASIS, ACM, WDSI, *Information Systems Educators Conference*, *The Informing Science Institute*, and the *International Conference on Information Quality* at MIT (twelve fast-tracked to peer-reviewed journals), four book chapters, and one scholarly book. He was a member of ACM, a charter member of the *Association for Information Systems*, and a founding member and honorary fellow of the *Institute of Informing Science*. He passed away on May 28, 2013.

Wojciech Gasparski is Professor of Humanities, DSc, Director and founder of the Business Ethics Centre (a joint unit of Kozminski University & Polish Academy of Sciences Institute of Philosophy and Sociology). He is the immediate past vice rector for research of the Kozminski University. He is professor emeritus of the Institute of Philosophy and Sociology, Polish Academy of Sciences (PAS) where he chaired the Academic Board and headed the Department of Praxiology as well as the Research Group for Ethics in Business and the Economy. His publications on human action theory (praxiology), business ethics, methodology, science studies, and systems theory are numerous. He is a member of the Warsaw Learned Society, Learned Society of Praxiology (Founder and Honorary President), Polish Philosophical Society, International Society of Business, Economics and Ethics, Academy of Management (USA), Society for Business Ethics, Science Studies Committee PAS (Honorary Member), Management Science Committee PAS. He was editor-in-chief of *Praxiology: The International Annual of Practical Philosophy and Methodology* (Transaction Publishers, USA). In 2006, he was nominated for the European Faculty Pioneer Award for Business in Society by Aspen Foundation (USA) and the European Academy of Business and Society.

He was awarded the Tadeusz Kotarbiński Medal by the Polish Academy of Sciences for his achievements in management science (2008).

Yasuhisa Hasegawa is Associate Professor of University of Tsukuba, Tsukuba, Japan. Until 2003, he was Research Associate of Nagoya University, Japan. Professor Hasegawa has published widely in the area of robotics, especially physically assistive robots for humans. His articles have appeared in the Advanced Robotics of RSJ and Conferences of Robotics and Automation Society, IEEE.

Grzegorz J. Nalepa, PhD, DSc (habilitation), holds a position of Assistant Professor in Jagiellonian University and AGH University of Science and Technology in Krakow, Poland. He has been an active researcher since 1999. His work is focused on design methods for intelligent systems, as well as knowledge representation and reasoning techniques. He formulated a new design approach for intelligent rule-based systems called XTT (eXtended Tabular Trees), as well as the Semantic Knowledge Engineering design and analysis methodology for knowledge-based systems. He authored and co-authored over 150 publications from the domains of rules and knowledge engineering, intelligent systems, semantic web, business intelligence, and software engineering. His fields of interest also include cognitive aspects of knowledge representation and reasoning in intelligent systems within the scope of cognitive science. Since 2012 his work is focused on two main application areas: business intelligence, including business rules and process modeling and analysis, and ambient intelligence including context aware applications for mobile devices. He has been involved in over thirty international conferences and workshops. He is the President of the Polish Artificial Intelligence Society (PSSI), member of ECAI.

Qiang Huang is Professor of Beijing Institute of Technology (BIT), the director of Intelligent Robotics Institute at BIT, and the director of Key Laboratory of Biomimetic Robots and Systems, Ministry of Education of China. From 1996 to 2000, he was Research Fellow in the Mechanical Engineering Laboratory (MEL), AIST and a researcher in the Department of Mechano-Informatics, University of Tokyo, Japan. His research interests include biped robots, humanoid robots, and intelligent mechatronics. His articles have appeared in the *IEEE Transactions on Robotics* and the *International Journal of Robotics Research.*

Pericle Salvini graduated in Foreign Languages and Literature from the University of Pisa and completed an MRes in Theatre Studies at Lancaster University (UK). In 2008, he received his PhD in Biorobotics Science and Engineering from IMT Lucca Institute for Advanced Studies and Scuola Superiore Sant'Anna (Italy). He is currently research fellow at the BioRobotics Institute of Scuola Superiore Sant'Anna, Pisa, Italy. His main research interests are in the fields of Human-Robot Interaction and Technoethics.

Kosuke Sekiyama is Associate Professor of Nagoya University, Nagoya, Japan. Until 2006, he was Associate Professor of University of Fukui, Fukui, Japan. Professor Sekiyama has published in the area of distributed autonomous systems, multiagents, ad hoc sensor networks. His articles have appeared in the IEEE Transactions, and Conferences of Robotics and Automation Society, IEEE.

Huma Shah is Research Fellow, University of Reading, England working on the EU FP7 Science in Society RoboLaw project. She is an expert in Turing's Imitation Game and coordinated the historic Turing100 event at Bletchley Park, on the hundredth anniversary of Alan Turing's birth in 2012. Her articles have appeared in *Kybernetes*, *Minds and Machines*, and UK House of Commons Science & Technology Committee's Report on Malware and Cybercrime.

Elettra Stradella is Assistant Professor in Comparative Public Law at the University of Pisa and Affiliate Researcher at the Scuola Superiore Sant'Anna of Pisa. She is author of many scientific publications in the field of Italian and Comparative Constitutional Law and collaborates with the RoboLaw project at Scuola Sant'Anna, investigating hard and soft law in regulating robotic technologies.

Krzysztof Tchoń, PhD, DSc, is Professor of Automation and Robotics at the Institute of Computer Engineering, Control and Robotics, Wrocław University of Technology, Poland, in charge of the Department of Fundamental Cybernetics and Robotics. His research interests concentrate on mathematical aspects of system theory and robotics as well as on social robotics. Professor K. Tchoń is a member of European Network for Social Intelligence and of the European research initiative RoboCom—Robot Companions for Citizens.

Ladislav Tondl, PhD, is Full Professor at the philosophical faculty Charles University in Prague. He studied philosophy, sociology, and economy at the same university. Earlier he was Associate Professor, and from 1968, has been a full professor. He was Director of the Institute for Theory and Methodology of Science at the Czechoslovak Academy 1968, 1969, President of Science Policy Commission UNESCO 1968–1970, Honourary Life member of the Association for the Philosophy of Science, USA, member of some learned societies abroad, and of editorial boards of some international journals and book editions. Because of political reasons, he was dismissed from Charles University after 1950, and after the invasion in 1968 from the Academy, he worked many years in the Institute of Information Theory and Automation in Prague engaged in the problems of logical semantics, scientific procedures, and the relationship between modern information theory and decision theory. After the Soviet invasion, he was dismissed from the Academy and worked in the sphere of computer-aided design and computer-graphics in a design organization. After his full rehabilitation in 1989, he has been Head of the Centre for Science Technology and Society Studies, Academy of Sciences Czech Republic.

Cristina Urdiales received her degree in Telecommunication Engineering from the Universidad Politecnica de Madrid (UPM) and PhDs from the Universidad de Malaga (UMA) and Universidad Politecnica de Catalunya (UPC) in Robotics and Artificial Intelligence, respectively. Currently she is working as Assistant Professor in UMA. Her research interests include autonomous motion, assistive technologies, computer vision, and Ambient Intelligence.

Anton Vedder, PhD, is an Associate Professor of Ethics and the Regulation of Technology and head of TILT, the Tilburg Institute for Law, Technology, and Society of Tilburg University, the Netherlands. His teaching and publication focus on ethics, regulation, and the adoption of innovative technologies.

Kevin Warwick is Professor of Cybernetics, University of Reading, England. He researches in Artificial Intelligence, Biomedical Engineering, Control Systems and Robotics; however, he is best known for his self-experimentation involving direct interfaces between the human nervous system and the Internet.

Hiroyuki Yoshikawa, Professor, is now Director General of Centre for Research and Development Strategy, Science and Technology Agency, Japan, where he is leading the design national research programmes for realizing the sustainable prosperity in society by integrating sciences: natural, medical, technological, social, and literal. He was President of University of Tokyo, Open University, AIST, ICSU, CIRP JSPS, and IMS. He established General Design Theory and developed theories of service and maintenance based on general design theory. He devoted his academic career to establish General Design Theory, which is based on topological model of human thought, common through different engineering disciplines. He led an international programme of collaborative research on manufacturing; IMS (Intelligent Manufacturing Program). He was President of ICSU, 1999–2003, and worked for its reformation.

For Product Safety Concerns and Information please contact our EU representative GPSR@taylorandfrancis.com
Taylor & Francis Verlag GmbH, Kaufingerstraße 24, 80331 München, Germany

www.ingramcontent.com/pod-product-compliance
Ingram Content Group UK Ltd.
Pitfield, Milton Keynes, MK11 3LW, UK
UKHW022119190926
475702UK00013B/1528
* 9 7 8 1 1 3 8 5 0 7 3 1 9 *